# Dragonflies and
# Damselflies
## of Britain and Western Europe
### A Photographic Guide

# Dragonflies and Damselflies

## of Britain and Western Europe
### A Photographic Guide

Authors: Jean-Pierre Boudot, Guillaume Doucet & Daniel Grand
Illustrations: Yves Doux

BLOOMSBURY WILDLIFE
LONDON · OXFORD · NEW YORK · NEW DELHI · SYDNEY

BLOOMSBURY WILDLIFE
Bloomsbury Publishing Plc
50 Bedford Square, London, WC1B 3DP, UK
29 Earlsfort Terrace, Dublin 2, Ireland

BLOOMSBURY, BLOOMSBURY WILDLIFE and the Diana logo are trademarks of
Bloomsbury Publishing Plc

First published in the United Kingdom by Bloomsbury Publishing 2021

This edition published by arrangement with Biotope editions, Meze, France

First published in France under the title *"Cahier d'identification des libellules
de France, Belgique, Luxembourg et Suisse" 2ème édition* by Biotope editions,
Meze. © Biotope editions, Meze, France 2019

Photos by Stéphane Hette, unless stated otherwise.

Cover photos © front cover: t Westend61/Getty Images, bl Westend61/Getty Images,
bc ullstein bild/Getty Images, br Marc Heath; spine: Jackie Bale/Getty Images; back cover:
tl Sandra Standbridge/Getty Images, tc David Sainsbury/Getty Images, tr Marc Heath.

A catalogue record for this book is available from the British Library.

Library of Congress Cataloguing-in-Publication data has been applied for.

ISBN: PB: 978-1-4729-8222-3
ePub: 978-1-4729-8221-6
ePDF: 978-1-4729-8220-9

2 4 6 8 10 9 7 5 3 1

Design by Rod Teasdale
Printed in Thailand by Cyberprint Group Co;Ltd

To find out more about our authors and books visit www.bloomsbury.com
and sign up for our newsletters.

# Contents

Azure Damselfly, *Coenagrion puella*

# Foreword

"Fresh water is key to life on Earth, is a core element of our climates and is a critical resource for the survival of humankind. Although freshwater habitats cover only 1 per cent of the Earth's surface, they are home to 10 per cent of the world's fauna. Sadly, however, they are often overlooked and the animal species that live there are more threatened than their counterparts living in other habitat types.

Dragonflies and damselflies are the ambassadors of the freshwater world. They have the extraordinary ability to disappear during bad weather, only to reappear in a wingbeat when the sun breaks through the clouds. No other animal embodies the beauty, fragility and resilience of the natural world better than dragonflies: we are transfixed by their penetrating gaze, we marvel at their agility and we are fascinated by their delicate bodies. Unsurprisingly, dragonflies are the subject of special attention from nature enthusiasts, and online biodiversity platforms receive tens of thousands of records of dragonfly observations each year. These data provide an opportunity to study the decline of certain species due to the destruction of their habitats, the increase of other species after hot summers or the return of others to restored rivers.

I would like to pay heartfelt tribute to Jean-Pierre Boudot, the leading specialist on European dragonflies over the past few decades. His most remarkable achievement has been the mapping of species, which he has been able to undertake through the close monitoring of new publications and by making long, solitary car journeys to survey sites around Europe. His work provided the basis for the *Atlas of the Odonata of the Mediterranean and North Africa* (2009), the *Atlas of the European Dragonflies and Damselflies* (2015) and the Red Lists of these regions, published in 2009 and 2010, respectively. It therefore comes as no surprise that a new species discovered in 2014 in Morocco was named *Onychogomphus boudoti* in his honour.

I had the great pleasure of visiting the Siwa Oasis with Jean-Pierre in 2009. On the edge of the Sahara Desert, close to the Libya–Egypt border, this isolated oasis is the only place in the world where the European dragonfly *Orthetrum coerulescens*, the Asian species *O. sabina* and the African species *O. machadoi* coexist and can be observed together. This site fascinated Jean-Pierre, because it offers a unique insight into how the climates of the Old World influenced the associations of dragonflies that can be observed today. Furthermore, the survival of these species in such a hostile environment is a metaphor for the nature of Jean-Pierre himself: a rare scientist who does not allow himself to be distracted by the latest methodological innovation or fashion, but who, through his stoicism and meticulous work, is accumulating a solid knowledge base on which everyone will be able to build."

Klaas-Douwe ('KD') B. Dijkstra
Research leader at Stellenbosch University.
Author of the *Field Guide to the Dragonflies of Britain and Europe.*

# 1

# The Life Cycle of Dragonflies and Damselflies

All dragonflies exhibit a similar life cycle, which can be divided into three main stages marked by a series of key events: **egg**, **larva** and **adult**. Together, they constitute a generation.

The number of generations per year varies with the species and, within the same species, with the climatic zone. Species that have only one generation per year are said to be **univoltines**; those that have more than one generation per year are called **multivoltines** (in our geographic region, these are bi- or trivoltines). Species with a longer larval cycle, spanning two years or more, are **semivoltines** (one generation every two years) or **partivoltines** (one generation every 3–6, or even 10 years).

## EGG-LAYING AND INCUBATION

Dragonflies employ many different egg-laying strategies. Some species (Zygoptera, Aeshnidae) use their ovipositor to inject their eggs into the stems of herbaceous vegetation, into the bark of softwood trees, into floating plant debris, or into submerged rotten wood or peat; this is called endophytic egg-laying. Other species (certain Anisoptera) carefully deposit their eggs on the surface of submerged or emerged vegetation, which serves as a simple support and to which they adhere; this is called epiphytic egg-laying. Some species drop their eggs at regular intervals above or at the surface of the water, or (Cordulegastridae) sink their eggs into the sediment at the bottom of streams using their ovipositor, by hovering vertically and stabbing their abdomen into the stream-bed; this is called exophytic egg-laying. The eggs of most species are cream to light brown in colour when they are laid.

In all cases, the female may either lay alone or remain in tandem with the male, the latter holding her by the prothorax or behind the eyes. The male may also accompany the female by flying in close attendance.

Endophytic egg-laying in *Calopteryx haemorrhoidalis*.

The duration of embryonic development is highly variable, depending on the species and environmental conditions. In some species, embryonic development begins immediately after the egg is laid and continues without interruption, so that hatching can occur before winter. In other species, embryogenesis is interrupted shortly after the egg is laid, and does not resume until after winter (winter diapause). Hatching is thus postponed until the following spring.

Exophytic egg-laying in *Cordulegaster boltonii*.

# The Odonata life cycle

➔ Male looking for a female ❶

➔ Mating ❷

➔ Egg-laying ❸

➔ Small larvae ❹

➔ Large larvae ❺

➔ Emergence ❻

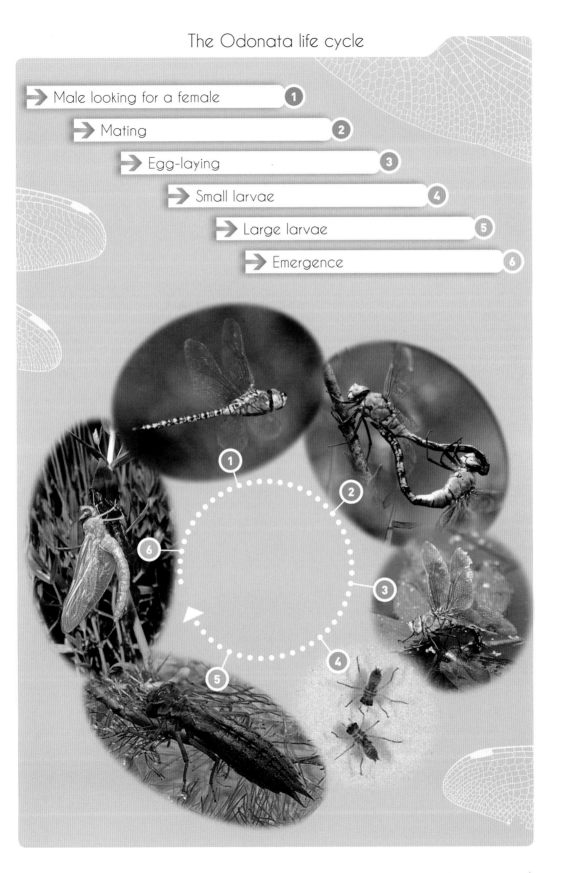

# THE LARVAL PHASE

Each egg hatches into a **prolarva** or **primary larva**. The larval phase is always aquatic in our regions and includes several stages. Each larval stage ends with a moult, also called **exuviation** or **ecdysis**. There are 8–18 moults, depending on the species, but the majority of species in both Zygoptera and Anisoptera go through a series of 11–13 moults.

## • ANATOMY OF LARVAE

During the different larval stages, the legs (initially reduced to one article) start to develop; the tarsi become specialised; the antennae acquire their final number of segments (4–7); the wing sheaths, absent at first, appear after the fifth moult and then develop; the genital organs are formed; the eyes grow; and, in Zygoptera, the caudal lamellae appear.

The **head** consists of the **labrum**, two **mandibles**, two **maxillae** and a **prehensile mask**. The main parts of the latter are the **submentum**, the **mentum** and two **labial palps**, each bearing a **mobile hook**. The mentum and the palps may be lined with **setae** (bristles).

The **thorax** bears **three pairs of legs** and the **wing sheaths**. On each side of the thorax are two pairs of respiratory stigmas (the meso- and metathoracic stigmas); these orifices and their internal extensions (tracheae and tracheoles) develop during successive moults and mark the progressive development of aerial respiration.

The **abdomen** is located behind the thorax and is made up of **10 segments**, these sometimes bearing spiny growths. The apex of the abdomen ends in an excretory orifice surmounted by appendages that differ in the Zygoptera and Anisoptera.

In the Zygoptera, the appendages are made of three well-developed **caudal lamellae**, whereas in the Anisoptera they are reduced to small pointed outgrowths that fit together to form the **anal pyramid**. The anal pyramid is a complex structure that includes two **cerci**, two **paraprocts** and an **epiproct**, the base of which is more or less widened.

## ANATOMY OF A LARVA (EXUVIA) OF ANISOPTERA

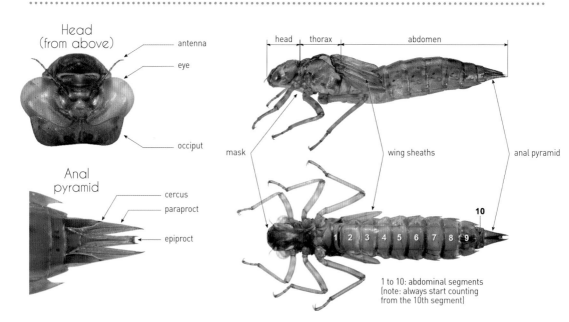

Head (from above) — antenna, eye, occiput

Anal pyramid — cercus, paraproct, epiproct

head · thorax · abdomen

mask · wing sheaths · anal pyramid

10

1 2 3 4 5 6 7 8 9

1 to 10: abdominal segments (note: always start counting from the 10th segment)

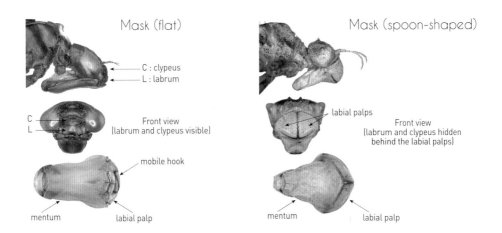

Mask (flat)

C : clypeus
L : labrum

Front view
(labrum and clypeus visible)

mobile hook

mentum          labial palp

Mask (spoon-shaped)

labial palps

Front view
(labrum and clypeus hidden
behind the labial palps)

mentum          labial palp

# ANATOMY OF A LARVA (EXUVIA) OF ZYGOPTERA

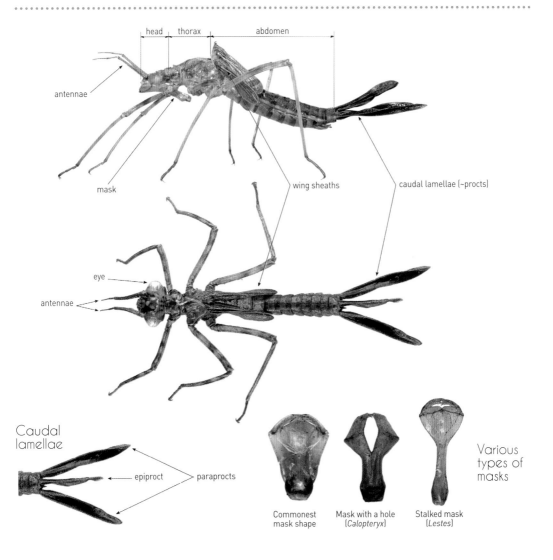

head | thorax | abdomen

antennae

mask

wing sheaths

caudal lamellae (~procts)

eye

antennae

## Caudal lamellae

epiproct          paraprocts

## Various types of masks

Commonest          Mask with a hole          Stalked mask
mask shape          (*Calopteryx*)          (*Lestes*)

## • BEHAVIOUR OF LARVAE

Zygoptera larvae are rather slender and agile, and are able to move by undulating their body. In contrast, Anisoptera larvae propel themselves rhythmically by vigorously ejecting water from their rectum. Usually stouter, the latter are often reluctant to move and can be seen walking slowly or burrowing into the substrate.

Dragonfly larvae are carnivorous and have predators. They also exhibit different types of posture. Some larvae, notably in the Libellulidae, live at the bottom of waterbodies, lying on substrates such as sediment or stones with their legs spread widely. Some blend into their environment, hiding in the surrounding mineral and organic debris. Others burrow into sediments, exposing only their head and upper chest on one side, and their anal pyramid on the other side (so that they can breathe). This behaviour is often encountered in Gomphidae and Cordulegastridae. Finally, some larvae do not come into contact with any mineral substrate and live on aquatic vegetation, with their legs folded around plant stems; several Zygoptera and many Aeshnidae display this behaviour.

## EMERGENCE

The larval phase ends with the larva exiting the water. At this final stage, the larva undergoes an **imaginal moult** or **emergence**, during which the adult is released from the larval body. The larval phase is often longer than the adult phase and can last 6–10 weeks (*Lestes dryas*, *Sympetrum danae*) to several years (for example, 5–6 years in *Cordulegaster* species at altitude). Extreme values reported in other parts of the world vary from 20 days to 10 years. In species adapted to desert and arid regions, the larval phase is very short, thus allowing the larva to complete its development before the environment dries up again (for example, 2–4 weeks in *Hemianax ephippiger*).

Preparations for metamorphosis begin a few days before emergence, with the absorption of branchial offshoots and histolysis of the muscles of the mask, which lose all functionality. The larva leaves the water and climbs up a fixed support at the edge of the water or several metres away.

The adult, or imago, emerges from the larval skin (exuvia) by hanging backwards from it or standing upright above it. After a long period of immobility, which allows the legs to dry and harden and the insect to recover, the adult spreads out its wings. It lets the wings dry and harden, after which it can finally take its first, or imaginal, flight. The metamorphosis ends a little after the first flight of the young imago.

Emergence usually takes place in the morning, but sometimes it happens at night. The duration of this stage is highly variable, depending on weather conditions and the species, but is usually 1–3 hours.

During this phase, the insect is defenceless and vulnerable to predation. The annual mortality rate at emergence appears to range from 3 per cent to 30 per cent, depending on the species and situation. Sometimes, daily mortality can reach 50 per cent of individuals that emerge.

# Emergence in Zygoptera: *Lestes barbarus*

# Emergence in Anisoptera: *Epitheca bimaculata*

# THE ADULT PHASE (IMAGO)

## ● ANATOMY OF ADULTS

As in other insects, the body of the adult consists of three well-defined parts: the **head**, the **thorax** and the **abdomen**.

The **head** is highly mobile and bears two large eyes that are either joined or separate. Dragonflies rely heavily on vision, and so their eyes are especially important features. At the top of the frons are two antennae.

The **thorax** consists of the prothorax (to which the head is attached), which carries the forelegs, and behind this, the **synthorax** (posterior part of the thorax), which carries the mid- and hindlegs, as well as the two pairs of wings.

The **two wing pairs** are similar in Zygoptera and dissimilar in Anisoptera, where the hindwings are much wider at the base than the forewings. Far more than any other trait, wing venation has helped to characterise dragonfly families. However, there has been an over-reliance on this approach, which has sometimes led to errors in the classification of dragonflies. Dragonflies, and Anisoptera in particular, are agile fliers. They can remain at a fixed point while hovering and then accelerate abruptly, can fly up or down vertically, and can fly backwards. While the flight of Zygoptera species is often more uncertain, they can occasionally display surprising agility. Flight performance is largely due to the independence of the forewings and hindwings, whose movements can be synchronous and simultaneous or, on the contrary, asynchronous.

The synthorax also bears two pairs of meso-metathoracic **stigmata**, which are used for breathing and are located above the attachment of the mid- and hindlegs. Other such stigmata exist on the underside of the abdomen, but they are very small and barely visible.

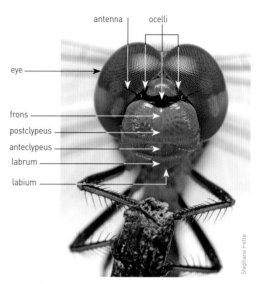

Close-up of the head of *Sympetrum striolatum.*

The **abdomen** is located behind the thorax and comprises 10 segments. It ends with the abdominal appendages, consisting of two **cercoids** (upper appendages) in all species and in both sexes. The lower appendages consist of two **cerci** in Zygoptera males, and a **supra-anal plate** in Anisoptera males. Females have no lower appendage. In males, the abdominal appendages are of particular importance, as they allow them to grip the females by the prothorax (Zygoptera) or by the head (Anisoptera) during mating. The configuration of the appendages is specific to each species and, generally, matches only the anatomy of the head or thorax of females of the same species.

The ventral surface of the second abdominal segment in males bears the **copulatory apparatus**, consisting of an articulated penis made of three articles and hamulis (hooks). Females that inject their eggs into plants (Zygoptera, Aeshnidae) have an **oviscapt** in line with segments 8 and 9, those that insert their eggs into sandy riverbeds (*Cordulegaster*) have an ovipositor, and those that drop their eggs into water have a **simple plate or vulval scale**.

## Wings and wing venation of Zygoptera **A** and Anisoptera **B**
[nomenclature from Tillyard & Fraser (1938–1940)]

The identification of dragonfly families relies mainly on traits linked to the wing venation. In this figure, we detail the nomenclature used in this book for wing traits.

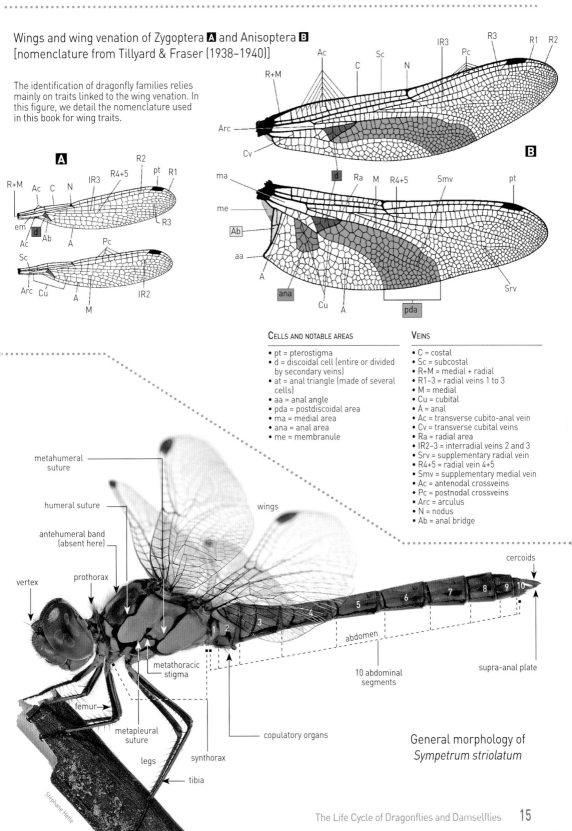

**CELLS AND NOTABLE AREAS**

- pt = pterostigma
- d = discoidal cell (entire or divided by secondary veins)
- at = anal triangle (made of several cells)
- aa = anal angle
- pda = postdiscoidal area
- ma = medial area
- ana = anal area
- me = membranule

**VEINS**

- C = costal
- Sc = subcostal
- R+M = medial + radial
- R1–3 = radial veins 1 to 3
- M = medial
- Cu = cubital
- A = anal
- Ac = transverse cubito-anal vein
- Cv = transverse cubital veins
- Ra = radial area
- IR2–3 = interradial veins 2 and 3
- Srv = supplementary radial vein
- R4+5 = radial vein 4+5
- Smv = supplementary medial vein
- Ac = antenodal crossveins
- Pc = postnodal crossveins
- Arc = arculus
- N = nodus
- Ab = anal bridge

General morphology of *Sympetrum striolatum*

Stéphane Hette

## Morphological variability

Structural polymorphism is exceptional in the Odonata. The only known case in Europe relates to females of *Boyeria irene*, which display two distinct morphological types. Some females, called *brachycerca*, have short cercoids (*c.* 2mm), while others, called *typica*, have much longer cercoids, measuring *c.* 6mm. Even if the former are the most numerous overall, the proportion of the two types is extremely variable and the *brachycerca* form can represent 40–100 per cent of the females present on a particular site, depending on the locality. These structural differences are clearly visible in exuviae and in larvae in their last stage. Within the same species, variations in size can commonly exceed 20 per cent. These variations may have a genetic origin, they may be caused by hormonal disorders or food deficiencies, or they may be caused by differences in habitat during the larval stages. They can also be linked to the time the adult emerges, which in turn is dependent on the timing of egg-laying.

## Sexual dimorphism

Dragonflies often exhibit **sexual dimorphism**, or differences between males and females of the same species. When the coloration of females differs from that of the males, the females are said to be **heteromorphic** or **gynochromic**. In this case, the males are adorned with much brighter colours than the females. For example, *Libellula depressa* males have a bright blue abdomen, while the females are yellow (this changes to yellow-brown, then blackish brown, with age, sometimes with a faint bluish pruinescence, but only in very old individuals).

In some cases, females display a similar or even identical coloration to that of males, in which case they are called **andromorphic** or **androchromic**. Such andromorphic females are frequent in certain *Calopteryx* species, and in several genera of Coenagrionidae (*Ceriagrion*, *Coenagrion*, *Enallagma*, *Ischnura*); they are much rarer in Aeshnidae and Libellulidae, but are well known in *Crocothemis erythraea* and *Sympetrum sanguineum*. A partial case of andromorphy is known for *Trithemis annulata*, where a female displayed a similar colour on the head as males but the rest of the body remained as in a normal female. Rare cases of andromorphism and partial androchromia have also been observed in *Libellula depressa*.

## Examples of polymorphism in females of *Ischnura elegans*

Heteromorphic female, *rufescens-obsoleta* morph, immature stage *rufescens*, pink.

Heteromorphic female, *rufescens-obsoleta* morph, young adult stage, orange.

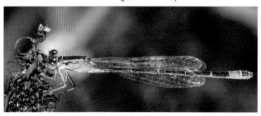

Heteromorphic female, *infuscans* morph, adult stage.

Purple immature stage, common to andromorphic and heteromorphic females of the *infuscans* morph.

## Chromatic variability

In some species, there may be great variability in the importance of melanism. For example, the spring generations of *Ischnura graellsii* are more melanic than the summer generations, especially on the thorax, where the pale antehumeral bands may disappear. These variations are significant, and are frequent in species such as *Coenagrion pulchellum* and *C. puella*.

The coloration of some species also changes with the age of the individuals, and large colour variations can be observed in the females of many dragonfly species. For example, in the genus *Ischnura* the adults change colour continuously during their maturation, and transient forms are known, with lighter parts that can be pink, orange, purple and/or blue.

# Change of coloration with age in males of *Orthetrum cancellatum*

One-day-old individual, yellow.

Maturing specimen.

Young mature adult.

Aged adult.

# • SEXUAL MATURATION

After emergence, the juvenile dragonfly sometimes moves away from water and begins a phase of sexual maturation, during which it will acquire its final colours. During this period most species seek resting grounds that are sheltered from the wind, and that heat up quickly in the sun, including forest clearings and edges, thickets, and sunken paths or those that are surrounded by hedges. In the Mediterranean region, certain species (for example, *Aeshna mixta* and *Sympetrum meridionale*) have a summer diapause during their physiological development, postponing their sexual maturation from the dry season, when aquatic environments are reduced, to the rainy season (autumn). When they reach maturity, most male Odonata return to an aquatic environment and await the arrival of females. Certain species appear to return en masse to the site of their birth, while others are seemingly able to adopt other sites – as long as these are suitable for breeding. In any case, movements between neighbouring environments are important, as shown by the rapid recolonisation of ponds that are regularly drained and refilled.

# • MATING

Mating in *Ceriagrion tenellum*.

As soon as a male notices a female, he will try to get hold of her. Mating can be preceded by rituals of presentation and identification, as in the genus *Calopteryx*. Here, the males hover in front of the females, bending their abdomen upwards. In other species, the sexual act is clearly less 'civilised', with the male grabbing hold of the female forcefully with his abdominal appendages.

The method by which the male grasps the female is different in the two Odonata suborders. In Zygoptera, the male's appendages are anchored on the top of the female's prothorax, as the structures are complementary. In Anisoptera, the male uses his three abdominal appendages to grasp the head of the female, between and behind the eyes. The female notifies the male that she accepts his offer to mate by bending her abdomen downwards. The male then brings the female back under him, so that the copulatory organs of the female, located under abdominal segments 8 and 9, can link with the male copulatory apparatus, located under the second abdominal segment. During mating, the bodies of the male and female take the shape of a heart, known as the copulatory heart. This position can last from a few seconds, as in Libellulidae (which mate in flight without landing), up to an hour or more in those species that copulate while perching on vegetation.

# • THE LIFESPAN OF DRAGONFLIES

The adult phase is entirely dedicated to the perpetuation of the species, and includes a period of maturation, a period of reproduction and a period of ageing, which ends in the death of the imago.

The lifespan of adult dragonflies is highly variable, depending on the species, meteorological factors and predation. The flight time of the various species is, in our region, less than a year. Only one genus (*Sympecma*, with three species in Eurasia, including two in our region) can hibernate in its adult state. In this genus, emergence takes place from June to August, and egg-laying from March to May; the imagos then die and are replaced by a new generation that will hibernate. The other species of dragonfly in our region emerge between the end of March or April and August, and then gradually disappear over the course of the year. Overall, the end of the breeding period for a species coincides with the decline of its populations. At that stage, we observe withered and ageing individuals that are worn out by bad weather, by flying through thickets and reedbeds, by combat with potential rivals and by breeding efforts. The last individuals are seen flying until December in the southern half of Europe and around the beginning of November further north. The average lifespan of adults that have survived the maturation phase is 3–24 days for Zygoptera (median: 8 days) and 6–38 days in Anisoptera (median: 11 days), due to predation and bad weather. The average overall lifespan is 15–77 days in Zygoptera (median: 30 days), and 17–64 days in Anisoptera (median: 39 days), at least in our region.

# 2

# Dragonflies and their Habitats

In our region, as in other parts of the world, drag-onflies are dependent on water to complete their life cycle. However, they inhabit a wide range of habitats. While certain species are ubiquitous (for example, *Aeshna cyanea* can breed in almost all types of aquatic environment, from lowlands to the upper limit of forests in mountain ranges), others are more specialised and have a stronger or less strong association with a particular type of habitat.

There is a gradual rather than an abrupt decrease in the number of native species with increasing altitude. However, certain taxa (for example, *Leucorrhinia pectoralis*) are really specific to lowland areas, and, in constant habitat, disappear above a certain altitude. Other, usually less numerous, species (for example, *Aeshna subarctica* and *Somatochlora alpestris*) are restricted to high-altitude areas, and are observed only above 700–800m in the temperate zone.

As with many animals and plants, it is possible to define groupings of dragonfly species according to habitat type. Dragonflies are not randomly distributed – for example, some species are restricted to flowing waters and others to stagnant waters (see Checklist, page 150). This affinity of a group of species for a particular type of habitat leads to the formation of suites, or sets of species found fairly consistently in the same type of habitat. A given species can be present in several suites, and in a given site not all the species in a suite are necessarily present.

In general, flowing waters are favoured by many species in the families Calopterygidae, Platycne-mididae, Gomphidae, Cordulegastridae and Mac-romiidae, or to certain species of Coenagrionidae, Aeshnidae and Corduliidae. Standing waters attract a majority of Lestidae, Coenagrionidae, Aeshnidae, Corduliidae and Libellulidae.

## STAGNANT WATERS

## FLOWING WATERS

Stagnant waters include pools, ponds, large lakes, marshes, and acidic or alkaline peatlands. Some of these habitats are permanent, while others are temporary. Most of them are freshwater habitats, although some are brackish. The suites found in stagnant waters represent 80 per cent of the Odonata species in our region.

Some species are adapted to many habitat types, while others are specialised and may breed more in brackish environments (*Lestes macrostigma*) or in acid or peaty waters (*Coenagrion hastulatum, Somatochlora arctica, Somatochlora alpestris, Sympetrum danae, Leucorrhinia dubia*). Habitat specialists rarely form large or sustainable populations in other types of environment.

Flowing waters form open ecosystems, whose fauna and flora change from upstream to downstream. Upstream areas are generally characterised by clear, fresh, fast-flowing and well-oxygenated waters. Downstream, the water takes in sediments, heats up and, when polluted by organic matter, can lose a large part of its oxygen.

The suites found in flowing waters are easier to define than those of stagnant waters, because the diversity of taxa is lower. Flowing waters harbour rather demanding species, which are able to find ecological niches adapted to their needs. There are several main types of suite found in flowing waters, each with multiple variations depending on local environmental parameters.

# Distribution of the Suites of Dragonfly Species, According to Various Habitats

## 1 PEATLANDS AND HIGH-ALTITUDE LAKES

Lestes sponsa
Lestes dryas
Enallagma cyathigerum
Pyrrhosoma nymphula
Coenagrion hastulatum
Coenagrion puella

Aeshna juncea
Aeshna cyanea
Aeshna grandis
Aeshna caerulea*
Cordulia aenea
Somatochlora metallica

Somatochlora alpestris
Somatochlora arctica
Libellula quadrimaculata
Sympetrum danae
Sympetrum flaveolum
Leucorrhinia dubia

## 2 PEATBOGS

Lestes sponsa
Lestes dryas
Enallagma cyathigerum
Pyrrhosoma nymphula
Coenagrion lunulatum*
Coenagrion hastulatum
Coenagrion puella
Aeshna grandis
Aeshna juncea
Aeshna subarctica
Aeshna cyanea
Somatochlora metallica
Somatochlora flavomaculata
Somatochlora alpestris
Somatochlora arctica
Cordulia aenea
Libellula quadrimaculata
Sympetrum danae
Sympetrum flaveolum
Leucorrhinia pectoralis
Leucorrhinia dubia
Leucorrhinia rubicunda*

## 3 HEADWATER AREAS AND BASIN HEADS

Calopteryx haemorrhoidalis
Calopteryx virgo
Coenagrion mercuriale
Boyeria irene
Onychogomphus uncatus

Cordulegaster bidentata
Cordulegaster boltonii
Orthetrum brunneum
Orthetrum coerulescens

## 4 PONDS AND LAKES

Sympecma fusca
Lestes virens
Lestes sponsa
Chalcolestes viridis
Platycnemis pennipes
Enallagma cyathigerum
Pyrrhosoma nymphula
Ceriagrion tenellum
Coenagrion scitulum
Coenagrion pulchellum
Coenagrion puella

Ischnura elegans
Ischnura genei
Erythromma najas
Erythromma viridulum
Aeshna isoceles
Aeshna grandis
Aeshna cyanea
Aeshna affinis
Aeshna mixta
Brachyton pratense
Anax parthenope

Anax imperator
Epitheca bimaculata
Cordulia aenea
Somatochlora metallica
Somatochlora flavomaculata
Crocothemis erythraea
Trithemis annulata
Libellula quadrimaculata
Libelulla depressa
Libellula fulva
Orthetrum albistylum

Orthetrum cancellatum
Orthetrum brunneum
Sympetrum sanguineum
Sympetrum flaveolum
Sympetrum meridionale
Sympetrum striolatum
Sympetrum vulgatum
Leucorrhinia caudalis
Leucorrhinia albifrons
Leucorrhinia pectoralis

## 5 POOLS

Lestes barbarus
Lestes virens
Lestes sponsa
Lestes dryas
Ischnura elegans
Ischnura pumilio
Ischnura genei
Pyrrhosoma nymphula
Coenagrion lunulatum*

Coenagrion scitulum
Coenagrion puella
Erythromma viridulum
Aeshna isoceles
Aeshna affinis
Aeshna cyanea
Brachytron pratense
Anax parthenope
Somatochlora flavomaculata

Crocothemis erythraea
Libellula quadrimaculata
Libellula depressa
Sympetrum pedemontanum
Sympetrum depressiusculum
Sympetrum fonscolombii
Sympetrum flaveolum
Sympetrum meridionale
Sympetrum striolatum

## 6 PIONEER SPECIES

Ischnura pumilio
Crocothemis erythraea
Libellula depressa
Sympetrum striolatum

## 7 STREAMS AND SMALL RIVERS

Calopteryx haemorrhoidalis
Calopteryx virgo
Calopteryx xanthostoma
Calopteryx splendens
Chalcolestes viridis
Platycnemis acutipennis
Platycnemis latipes
Ceriagrion tenellum
Coenagrion mercuriale
Coenagrion ornatum*
Coenagrion caerulescens*

Erythromma lindenii
Boyeria irene
Gomphus vulgatissimus
Ophiogomphus cecilia
Onychogomphus uncatus
Onychogomphus forcipatus
Trithemis annulata
Orthetrum brunneum
Orthetrum coerulescens
Cordulegaster boltonii

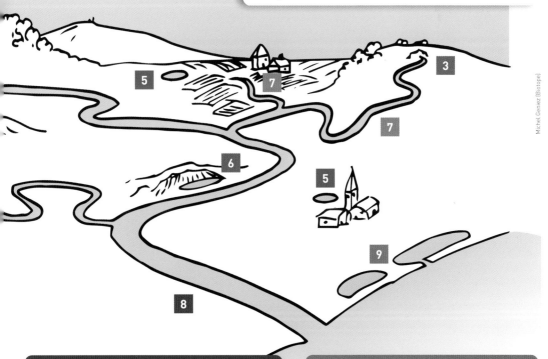

Michel Geniez (Biotope)

## 8 LARGE RIVERS

Calopteryx xanthostoma
Calopteryx splendens
Platycnemis acutipennis
Platycnemis latipes
Platycnemis pennipes
Boyeria irene
Ophiogomphus cecilia
Gomphus vulgatissimus

Gomphus graslinii
Gomphus pulchellus
Gomphus simillimus
Stylurus flavipes
Onychogomphus uncatus
Onychogomphus forcipatus
Macromia splendens*
Oxygastra curtisii

## 9 BRACKISH WATERS

Lestes barbarus
Lestes dryas
Lestes macrostigma
Ischnura pumilio
Ischnura elegans

Aeshna isoceles
Anax parthenope
Crocothemis erythraea
Sympetrum fonscolombii
Sympetrum meridionale

* localised species, restricted to this habitat

Norfolk Hawker, *Aeshna isoceles*

Scarce Chaser, *Libellula fulva*

# Dragonfly Surveying, Observation and Photography

## SURVEYING DRAGONFLIES

It is essential to survey the species inhabiting wetland habitats in order to guarantee their conservation. When habitats are found to be particularly remarkable, they can be classified as protected areas or managed by environmental protection organisations. Carrying out surveys and monitoring requires considerable human resources, which go far beyond the reach of public authorities or professional scientists. Such activities therefore remain the domain of amateur naturalists, whether they act individually or on behalf of specialist societies. Involvement with a society is always recommended for amateur entomologists, because it facilitates the transmission of information and its use for conservation purposes.

## OBSERVING DRAGONFLIES

Dragonflies can be observed in wetlands for two-thirds to three-quarters of the year in lowland areas, and more specifically in summer in mountainous regions. Spending some time at the end of spring near a meadow stream with well-vegetated banks or around a sunny pond sheltered from the wind will give you the opportunity to encounter a large number of species, as individuals are concentrated at the edge of the water at this time. The best period to observe dragonflies is from the beginning of May to the beginning of October in lowland areas, or in high summer at higher altitudes and latitudes. In lowlands, the hottest weeks of July and August usually lead to a decrease in dragonfly numbers. Autumnal species will then appear, ready to begin their breeding phase. Almost any wet habitat can be used by dragonflies as a breeding site – even the smallest, the saltiest and sometimes the most polluted sites! However, it would be wrong to believe that the quality of the water has no effect on the populations of Odonata, because egg-laying does not imply successful reproduction. We have seen *Anax imperator* lay eggs in a boat moored on the Adriatic Sea and *Cordulegaster* lay eggs on wet roads – these individuals are wasting their time and their energy! Remember also that given species do not just live anywhere, so if you are looking for a particular species, you will have to explore its preferred habitats.

When surveying for dragonflies, avoid wearing clothes that are light in colour or contrast strongly with the surrounding environment, as they will make your movements more noticeable. In addition to appropriate clothing – and wellies if you want to avoid getting your feet wet – you will need various other items to improve your efficiency during any survey of Odonata.

As a start, your basic equipment should include an insect net (preferably one that can be taken apart), with an adjustable telescopic handle 1–2m or more in length (some models extend to 5m). The net should have a foldable hoop (diameter 40cm) and green or brown mesh. In addition, your survey kit should include a hand lens with 10× magnification (called a geology loupe), and either a notebook, a dictaphone or a mobile tablet. All the information collected at the observation sites must be recorded: locality, altitude, weather conditions, characteristics of the site, and the species observed, their abundance and their behaviour. This basic equipment may be supplemented by a pair of binoculars or a monocular, a local topographic map and a GPS to determine the geographic coordinates of the observations.

The best way to remove a captured individual from your net is to grab it gently by the thorax or wings. Except in species that are difficult to identify and require examination under a binocular microscope or a microscope, the identification may take place *in situ*, the insect then being released on the spot. Never catch juveniles: doing so can cause irreversible damage, as their tissues are not yet fully sclerotised and cannot bear any contact.

Identification keys must be used rigorously, first to determine the family, then the genus and finally the species or subspecies.

Beginners should keep in mind that females are often more difficult to identify than males, especially in some genera such as *Coenagrion*.

Regarding these difficult groups, beginners should try, at least during their first years of surveying, to base their species list solely on the identification of males.

# PHOTOGRAPHING DRAGONFLIES

The practice of insect macrophotography is widespread among nature enthusiasts. It requires a little technique and a lot of patience, but, importantly, it allows you to back up your observations without the need for insect collection. When taking photographs for identification purposes, it is therefore important to highlight traits that are useful for distinguishing species. You can also send any photographs you take to recorders, so that your observations can be confirmed and added to local databases. In addition, photography makes it possible to compile documentation regarding dragonfly behaviour (mating, egg-laying, capture of prey and so on), which is impossible through the collection of specimens. Taking photographs is also the only way to confirm the presence of a protected species, where capture of such a species is illegal. The photography of species can further be complemented by habitat photography, and the images taken can then be indexed in relation to your field notes. In the case of species found outside their known range, a photograph georeferenced using the camera's GPS and saved in RAW format allows for the discovery to be validated and eliminates any risk of fraud.

As dragonflies are naturally suspicious, photographing them is often a game of patience. A reflex (DSLR) camera fitted with a macro lens with a focal length of 100–180mm (to which a TTL flash may be added if the light conditions are not optimal) will enable you to photograph dragonflies in most situations. With current high-end digital cameras, it is easy to operate under high sensitivity at sufficient speed and at medium aperture, while avoiding any noise.

The use of a zoom lens is not recommended, because the optical quality of such lenses is generally quite low for close-up shots.

# BEST PRACTICE

During all surveying and monitoring activities it is important to act within the rule of law, keeping in mind that species and localities may be protected. Observation and any ancillary activities must therefore always take place in strict respect of the properties and habitats visited, avoiding damage and excessive trampling (wetlands being particularly fragile in this respect). The habitat should be left in the state in which it was found. Many dragonfly sites are situated on private property, where permission may be required to access them. From our own experience, refusals to such requests are relatively rare, and the real difficulty instead lies in identifying and contacting the owner. Access to localities classified as nature reserves is often regulated or prohibited outside of specific scientific studies. Once again, it is best to contact the organisation that manages the site.

Entomological observation or survey activities sometimes lead you to come into contact with protected species. In these cases, damage to habitat, egg collection, and capture of larvae or adults are all prohibited without prior authorisation.

In the presence of protected species and the absence of the necessary authorisation to capture them, it is nevertheless a pleasure to observe these insects in their natural habitat and, when possible, to record this with photographs or to collect their exuviae.

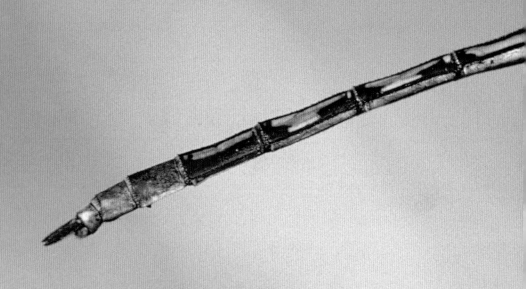

# Field Key to
# Adult Dragonflies

Long Skimmer, *Orthetrum trinacria*

Alain Cochet

# HOW TO USE THIS KEY

**1** Name of family or genus
**2** Distribution map
**3** Flight period: JFMAMJJASOND
**4** Distribution and abundance
**5** Habitat

**6** Confusion species
**7** Length of abdomen (including abdominal appendages)
**8** Scientific and common names
**9** Additional identification criteria

**A** Emergence and start of the breeding period
**B** Main breeding period
**C** Decline of adult individuals

| Dichotomous key | Drawings and diagrams | Distribution and habitat | Photographic plate ♂♀ |

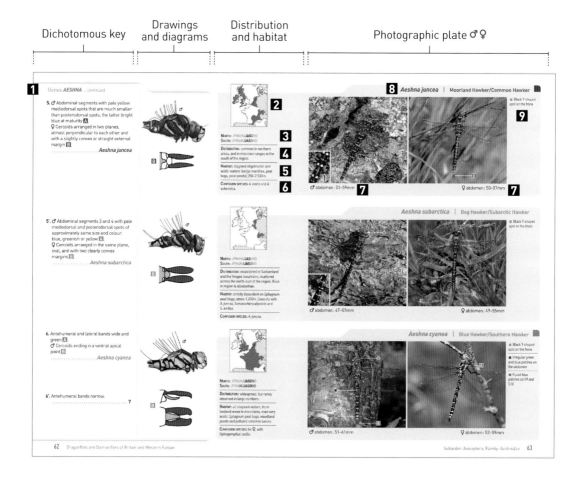

# KEY TO FAMILIES

Base of wings with 2 transverse antenodal veins.
Wings stalked, rather narrow and elongated, with a true pterostigma in both sexes.

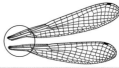

Forewings and hindwings similar.
At rest, the wings are held shut (sometimes slightly open).
.................Suborder **Zygoptera**

Base of wings with more than 2 transverse antenodal veins.
Wings unstalked, rather wide, more or less opaque, coloured or hyaline.
Pseudopterostigmas (pterostigmas crossed by veins), only present in females.
.................**Calopterygidae** (p. 32)
(1 genus: *Calopteryx*)

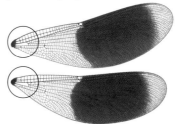

**O**DONATA

Forewings narrower than the hindwings.
At rest, the wings are spread out.
........ Suborder **Anisoptera**

Eyes distinctly separated.
...................**Gomphidae** (p. 68)
(6 genera: *Gomphus, Stylurus, Paragomphus, Onychogomphus, Ophiogomphus, Lindenia*)

Eyes only just touching, or eyes joined together.

Pterostigmas covering 2–4 cells.
. . . . . . . . . . . . . . . . . . . **Lestidae** (p. 36)
(3 genera: *Lestes*, *Chalcolestes*, *Sympecma*)

Pterostigmas covering 1–1.5 cells.

Eyes distinctly joined together over a certain length.
. . . . . . . . . . . . . . . . . . **Aeshnidae** (p. 58)
(5 genera: *Boyeria*, *Anax*, *Hemianax*, *Brachytron*, *Aeshna*)

Eyes coming into contact in a defined spot, or almost touching.

Discoidal cells distinctly trapezoidal. Tibiae not widened. Head only slightly widened transversally; upper surface of head black, marked with postocular spots or with a pale line.
. . . . . . . . . . . . . . . . . . . . . . . . . . . . . **Coenagrionidae** (p. 44)
(7 genera: *Coenagrion*, *Pyrrhosoma*, *Ischnura*, *Erythromma*, *Nehalennia*, *Ceriagrion*, *Enallagma*)

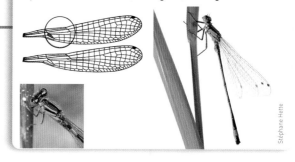

Stéphane Hette

Discoidal cells almost rectangular.
Tibiae widened, at least in males.
Head strongly widened transversally, with two pale lines surrounding a dark band (except in melanic forms).
. . . . . **Platycnemididae** (only 1 genus: *Platycnemis*) (p. 42)

Stéphane Hette

Discoidal cells longitudinal on all four wings (in the same direction).
. . . . . . . . . . **Cordulegastridae** (1 genus: *Cordulegaster*) (p. 76)

Jérôme Robin (Biotope)

Discoidal cells in opposite directions: transverse on forewings, and longitudinal on hindwings.

3–4 transverse cubitoanal veins between the discoidal cell and the base of the wing. Presence of a square-rounded anal area on hindwings.
. . . . . . . . . . . . . . . . . . . . . **Macromiidae** (p. 76)
(only 1 species: *Macromia splendens*)

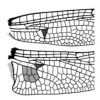

Posterior edge of the eyes without marked sinuosity.
No anal angle on male hindwings.
. . . . . . . . . . . . . . . . . . . . . .**Libellulidae** (p. 82)
(9 genera: *Libellula, Orthetrum, Crocothemis, Sympetrum, Leucorrhinia, Trithemis, Selysiothemis, Brachythemis, Pantala*)

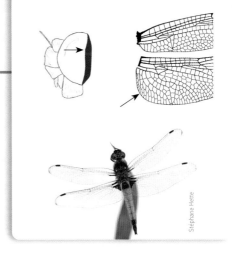

Stéphane Hette

1 or 2 transverse cubitoanal veins between the discoidal cell and the base of the wing. Presence of an elongated anal area on hindwings.

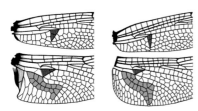

Posterior edge of the eyes with a marked sinuosity.
Presence of an anal angle on male hindwings.
. . .**Corduliidae and *Oxygastra curtisii*** (p. 78)
(4 genera: *Cordulia, Somatochlora, Oxygastra, Epitheca*)

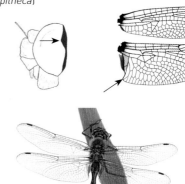

Stéphane Hette

# Family Calopterygidae

1. ♂ **Body brilliant reddish brown to metallic purple-brown** A. Underside of last three abdominal segments uniform, bright, carmine pink B. Wings suffused with a brown or dark brown tint at maturity C C C". ♀ Wings smoky, with a **dark brown area towards apex of hindwings** D. Body bright brown to metallic green, with a more or less coppery tint, fading with age.
........*Calopteryx haemorrhoidalis*

1'. ♂ **Body metallic blue or green at maturity** A. Underside of last three abdominal segments lack carmine-red coloration. Wings partly suffused with green or dark blue at maturity.
♀ Wings usually translucent or more or less smoky.
.................................. **2**

2. ♂ **Ventral surface of the last three abdominal segments reddish to dull orange** B B. Wings wide (ratio of forewing length:width = 2.4–3) and entirely blue-green, or with a paler coloration at apex or base. Coloration begins at costal vein, well before nodus C C.
♀ Pseudopterostigmas rather distant from apex (ratio of nodus–pseudopterostigma distance:pseudopterostigma–apex distance = 2.7–4) D D.
.................... ***Calopteryx virgo***
(two subspecies: ***Calopteryx virgo virgo*** and ***Calopteryx virgo meridionalis***)

• Wings of males translucent at base, for less than 5mm. **Apex usually translucent for 1–4mm**, especially on hindwings, sometimes opaque C. In females, ventral side of synthorax black with small yellow lines. Wings of females often slightly smoky D.
.............. *Calopteryx virgo virgo*

• Wings of males translucent from base to discoidal cell, or a little beyond it, for more than 5mm. **Apex of wings always coloured** C. In females, anterior two-thirds of ventral surface of synthorax mainly yellow. Wings of females sometimes very smoky, occasionally with a spot or brown bar towards apex D.
...... *Calopteryx virgo meridionalis*

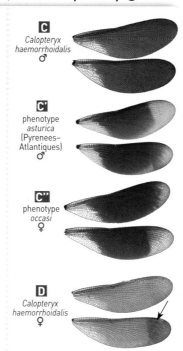

C
*Calopteryx haemorrhoidalis*
♂

C'
phenotype *asturica* (Pyrenees–Atlantiques)
♂

C"
phenotype *occasi*
♀

D
*Calopteryx haemorrhoidalis*
♀

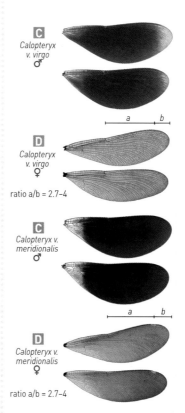

C
*Calopteryx v. virgo*
♂

D
*Calopteryx v. virgo*
♀
ratio a/b = 2.7–4

C
*Calopteryx v. meridionalis*
♂

D
*Calopteryx v. meridionalis*
♀
ratio a/b = 2.7–4

**NORTH AND SOUTH:** JFMAMJJASOND

**DISTRIBUTION:** generally common in th western Mediterranean; rarer in the north of its range.

**HABITAT:** clear, clean, well-oxygenated streams with relatively fast-flowing waters; below 1,100m.

**CONFUSION SPECIES:** for ♀, with those c *C. virgo meridionalis*.

**NORTH:** JFMAMJJASOND
**SOUTH:** JFMAMJJASOND

**DISTRIBUTION:** locally common across the region, although absent from larg areas of the south.

**HABITAT:** partly sunny, flowing waters, with a preference for hilly or mountainous areas; up to 1,600m.

**CONFUSION SPECIES:** for ♀, with those o *C. splendens*.

**NORTH:** JFMAMJJASOND
**SOUTH:** JFMAMJJASOND

**DISTRIBUTION:** common in south-weste Europe, even in mountainous areas. T two subspecies can coexist.

**HABITAT:** partly sunny, flowing waters.

**CONFUSION SPECIES:** for ♂, with those of *C. xanthostoma*; for ♀, with those of *C. splendens* and *C. xanthostoma*, and sometimes also with *C. haemorrhoidal*

## *Calopteryx haemorrhoidalis* | Copper Demoiselle

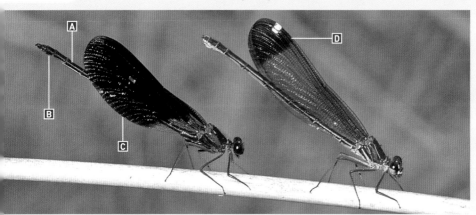

♂ abdomen: 34–43mm                    ♀ abdomen: 31–43mm

## *Calopteryx virgo virgo* | Beautiful Demoiselle

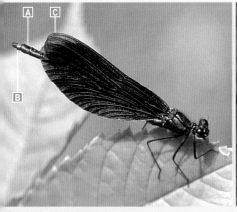

♂ abdomen: 33–42mm                    ♀ abdomen: 31–41mm

## *Calopteryx virgo meridionalis* | (Southern) Beautiful Demoiselle

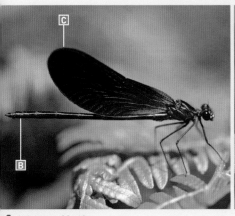

♂ abdomen: 33–42mm                    ♀ abdomen: 31–41mm

**2'.** ♂ **Ventral surface of last three abdominal segments yellow to greyish white** 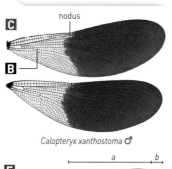 A A A.
Wings narrow (ratio of forewing length:width = 3–3.7) **B**. Coloration begins at costal vein, at nodus, and extends to base of wing **C C C**.
♀ Pseudopterostigmas fairly close to apex of wings (ratio of nodus–pseudopterostigma distance:pseudopterostigma–apex distance = 3.4–8.1).
. . . . . . . . . . . . . . . . . . . . . . . . . . . . . . **3**

**3.** ♂ **No translucent area at apex of wings C**.
Coloration appears a few days after emergence.
♀ Cercoids yellowish with a brown apex **D**. Ratio of nodus–pseudopterostigma distance:pseudopterostigma–apex distance = 6.7–8.1 **E**.
. . . . . . . . . . . .*Calopteryx xanthostoma*

*nodus*

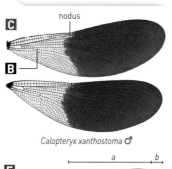

*Calopteryx xanthostoma* ♂

a    b

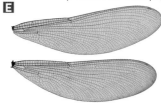

*Calopteryx xanthostoma* ♀
ratio a/b = 6.7–8.1

**NORTH AND SOUTH :** JFMAMJJASO

**DISTRIBUTION:** common in southern France and the Iberian Peninsula, becoming scarcer in the latter.

**HABITAT:** sunny southern streams with a moderate current and floating aqua plants; up to 1,200m in mountainous areas.

**CONFUSION SPECIES:** for ♂, with *C. virgo meridionalis*; for ♀, with *C. splendens*, *C. xanthostoma* and *C. virgo*.

**3'.** ♂ **Wide translucent area (1–7mm) at apex of wings.** Coloration visible immediately after emergence.
♀ Cercoids black **D**. Ratio of nodus–pseudopterostigma distance: pseudopterostigma–apex distance = 3.4–6.3 **E E**.
. . . . . . . . . . . . . . *Calopteryx splendens*
(two subspecies: *Calopteryx splendens splendens* and *Calopteryx splendens caprai*)

• Translucent area at tip of wings of males generally quite wide (1–7mm), but individuals with only a narrow transparent apical margin are frequent **C**.
. . . .*Calopteryx splendens splendens*

• Translucent margin at tips of wings in males generally always narrow (1–2mm) **C**.
. . . . . . . .*Calopteryx splendens caprai*

*nodus*

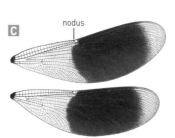

*Calopteryx splendens splendens* ♂
(not present in Corsica)

*nodus*

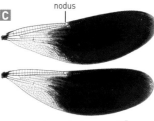

*Calopteryx splendens caprai* ♂
(Corsica and Southern Switzerland)

a    b

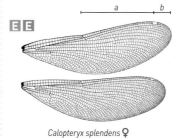

*Calopteryx splendens* ♀
ratio a/b = 3.4–6.3

**NORTH :** JFMAMJJASOND
**SOUTH :** JFMAMJJASOND

**DISTRIBUTION:** common at low and mid-altitudes except in the Scottish Highlands and Fennoscandia.

**HABITAT:** sunny, flowing waters at low and mid-altitudes; sometimes up to 1,200m. Occasionally observed in ponds and backwaters fed by groundwater.

**CONFUSION SPECIES:** for ♂, with *C. xanthostoma*; for ♀, with *C. splendens*, *C. xanthostoma* and *C. virgo*.

**NORTH AND SOUTH :** JFMAMJJASOND

**DISTRIBUTION:** Corsica and Switzerland, where it remains quite rare.

**HABITAT:** sunny, flowing waters.

**CONFUSION SPECIES:** for ♂, with *C. xanthostoma*; for ♀, with *C. splendens*, *C. xanthostoma* and *C. virgo*.

**CONFUSION SPECIES:** *C. s. caprai* males that have only a thin translucent border at the apex of their wings can easily be confused with *C. xanthostoma*, but their ranges do not overlap in our region. *C. splendens*, *C. xanthostoma* and *C. virgo* females can all be confused. Hybrids between *C. splendens sensu lato* and *C. xanthostoma* are known from France and Italy, and have often been confused with *C. s. caprai*, which is genetically distinct.

## *Calopteryx xanthostoma* | Western Demoiselle/Yellow-tailed Demoiselle

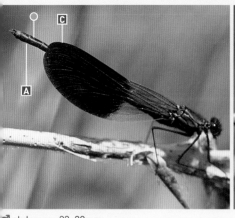

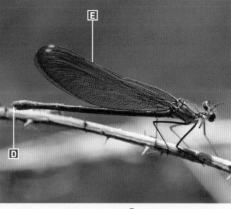

● Body blue to metallic green

♂ abdomen: 33–39mm

♀ abdomen: 33–39mm

## *Calopteryx splendens splendens* | Banded Demoiselle

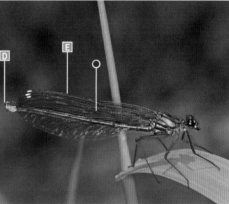

🅰 Underside of the last three abdominal segments of males yellow to greyish white, with black markings

● Body blue to metallic green

● Wings of heteromorphic females translucent to slightly smoky

♂ abdomen: 33–41mm

♀ abdomen: 33–40mm

## *Calopteryx splendens caprai* | (Capra's) Banded Demoiselle

● Body blue to metallic green

● Wings of heteromorphic females translucent to slightly smoky

♂ abdomen: 33–41mm

♀ abdomen: 33–40mm

# Family Lestidae

**1.** Discoidal cells on forewings significantly shorter than those on hindwings 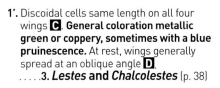. **Body dull brown** (but with a weak metallic sheen in young stages). At rest, wings held together over back **B**. Pterostigmas brown, elongated; that of hindwing offset from that of forewing **C** **C**.
.............**2. *Sympecma*** (see below)

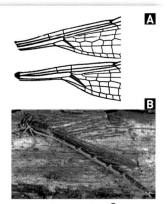

*Sympecma fusca* ♀

**1'.** Discoidal cells same length on all four wings **C**. **General coloration metallic green or coppery, sometimes with a blue pruinescence.** At rest, wings generally spread at an oblique angle **D**.
.....**3. *Lestes* and *Chalcolestes*** (p. 38)

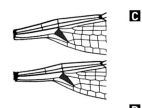

*Lestes dryas* ♂

## Genus *SYMPECMA*

**2.** Dorsal surface of thorax dark brown, with a **straight lower margin** in both sexes **A**. Cerci of ♂ reaching or extending beyond end of inner basal tooth of cercoids **B**.
.................... *Sympecma fusca*

**2'.** Dorsal surface of thorax dark brown, **lower margin with a distinct bulge** in both sexes **A**. Cerci of ♂ not reaching end of inner basal tooth of cercoids **B**.
.............. *Sympecma paedisca*

**B**
cerci

cercoids

**B**
cerci

cercoids

NORTH: JFMAMJJASOND
SOUTH: JFMAMJJASOND

DISTRIBUTION: generally well distributed in lowland plains across Europe. Absent from Britain, Ireland and most of Fennoscandia, but expanding northwards.

HABITAT: standing waters, even sometimes brackish waters; can be found above 1,100m.

CONFUSION SPECIES: *S. paedisca*.

NORTH AND SOUTH: JFMAMJJASOND

DISTRIBUTION: very localised, in Poland, north Germany and the Netherlands, and in the foothills of the Alps.

HABITAT: peat bogs with *Sphagnum* mosses; oligotrophic and mesotrophic marshes, even temporary ones (but in water from May to August); ponds and lakes surrounded by sedge meadows or reedbeds, often in forested environments; old flooded gravel pits; up to 600m.

CONFUSION SPECIES: *S. fusca*.

● Body pale brown to coppery in juvenile stages (summer and autumn), dull brown during the breeding period (spring)

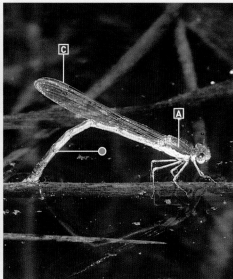

♂ abdomen: 26–31mm

♀ abdomen: 26–31mm

*Sympecma paedisca* | Siberian Winter Damselfly

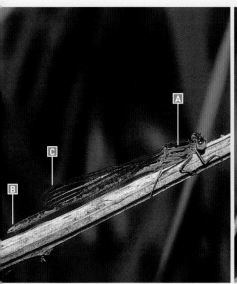

♂ abdomen: 25–31mm

♀ abdomen: 25–31mm

## Genera *LESTES* and *CHALCOLESTES*

3. **Occiput bicoloured, green** 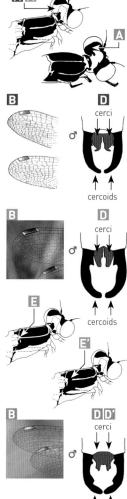 or coppery in upper half and pale yellow in lower half . . . . . . . . . . . . . . **4**

3'. **Occiput uniformly coloured** , metallic green to coppery, sometimes with additional blue pruinescence . . . . . . . . . . . . . . . . . . . . . . . . . . . . . **5**

4. **Pterostigmas clearly bicoloured**, with a brown base and whitish tip . Body metallic green to brilliant coppery brown, lacking blue pruinescence . Cerci of ♂ with a pointed tip, bent outwards .
. . . . . . . . . . . . . . . . . . . . . . . . . . . . *Lestes barbarus*

4'. **Pterostigmas uniformly brown**, or at most fading in their apical third . Body metallic green with a late blue pruinescence on abdominal segments 1 and 9–10, and on flanks of thorax in ♂ . Cerci of ♂ straight and with a rounded tip .
. . . . . . . . . . . . . . . . . . . . . . . . . . . . . *Lestes virens*
(two subspecies: *Lestes virens virens* and *Lestes virens vestalis*)

• Yellow dorsolateral stripe of humeral sutures wide, reaching base of hindwings .
. . . . . . . . . . . . . . . . . . . . *Lestes virens virens*

• Yellow dorsolateral stripe of humeral sutures always interrupted before base of wing in males, sometimes less markedly in females, occasionally narrow or absent .
. . . . . . . . . . . . . . . . . . . . *Lestes virens vestalis*

5. **Pterostigmas yellow to yellow-brown**, surrounded by dark veins . Body metallic green, lacking blue pruinescence . In dorsal view, cerci of ♂ angular and obliquely truncated at tip, . . . . . . . . . . . **6**

5'. **Pterostigmas dark brown or black** .
Well-marked blue pruinescence on certain parts of thorax and abdomen at maturity, at least in ♂. Cerci of ♂ with a rounded tip . . . . . . . . . . . . . . . . . . . . **7**

6. Internal lateral edge of ♂ cercoids with two strong lateral teeth; apical tooth thickened and more or less equilateral . In side view, cerci show a blunt tip. Ovipositor of ♀ with 10–14 ventral teeth at tip.
. . . . . . . . . . . . . . . . . . . . . . . *Chalcolestes viridis*

6'. Internal lateral edge of cercoids of ♂ with two small lateral teeth, apical tooth longer than wide . In side view, cerci show a strongly acute tip. Ovipositor of ♀ with 6–8 ventral teeth at tip.
. . . . . . . . . . . . . . . . . . . . . *Chalcolestes parvidens*

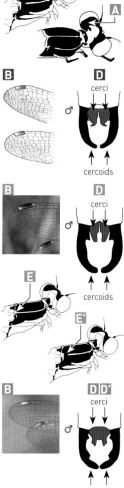

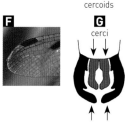

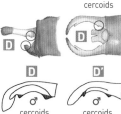

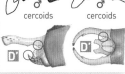

NORTH: JFMAMJJASOND
SOUTH: JFMAMJJASOND

**DISTRIBUTION:** common in the Mediterranean and on the Atlantic coa currently expanding northward. First recorded in Britain in 2002.

**HABITAT:** sunny, stagnant, shallow, unpolluted waters; sometimes even brackish and temporary waterbodies.

**CONFUSION SPECIES:** *L. virens*, whose pterostigmas have a whitish tip.

*Lestes virens virens*    *Lestes virens vest*

NORTH: JFMAMJJASOND
SOUTH: JFMAMJJASOND

**DISTRIBUTION:** *L. v. virens*, western Mediterranean, including Spain, southern France and Sardinia; scatter and generally infrequent. *L. v. vestalis*, common in northern France, Italy and central Europe; has become rare in Belgium and Luxembourg.

**HABITAT:** standing water; up to 1,400m

**CONFUSION SPECIES:** *Chalcolestes viridis* also has short cerci in young stages.

*Chalcolestes viridis*    *Chalcolestes parvi*

NORTH: JFMAMJJASOND
SOUTH: JFMAMJJASOND

**DISTRIBUTION:** *C. viridis*, common acros most of our region; absent from Britai except in the south-east, from most of Fennoscandia and from southern Gree *C. parvidens*, Corsica, where it is very rare (five known localities), and Italy, th Balkans and Turkey.

**HABITAT:** running and stagnant waters bordered by softwood trees; up to 1,50

**CONFUSION SPECIES:** *Lestes virens*, but th latter has a bicoloured occiput when you

## *Lestes barbarus* | Migrant Spreadwing/Southern Emerald Damselfly

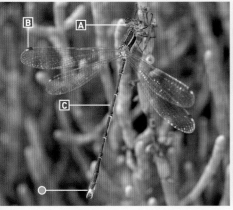

● In older males, weak white pulverulence at the tip of the abdomen

♂ abdomen : 26–35mm

♀ abdomen : 26–33mm

## *Lestes virens virens* | Small Spreadwing/Small Emerald Damselfly

♂ abdomen : 24–34mm

♀ abdomen : 24–30mm

## *Chalcolestes viridis* | Western Willow Spreadwing/Willow Emerald Damselfly

## *Chalcolestes parvidens* | Eastern Willow Spreadwing

● Cercoids white to yellow-brown

♂ abdomen : 33–40mm

♀ abdomen : 30–39mm

**7. Pterostigmas long and dark, usually spanning 3–4 cells**, rarely over only two 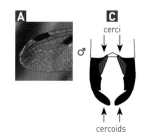. Thorax with a blue pruinescence, even on its dorsal side and in both sexes **B**. Cerci of ♂ short, wide and divergent **C**.
................*Lestes macrostigma*

**NORTH AND SOUTH:** JFMAM**JJASO**ND

**DISTRIBUTION:** mainly coastal and localised across southern Europe; also inland, especially on the Great Hungarian Plain.

**HABITAT:** brackish water; very rarely in fresh water (ponds, temporary ponds, marshes, ditches and coastal lagoons with bulrushes, rushes and sedges, bordered by saltflats with glassworts).

**CONFUSION SPECIES:** *L. sponsa* and *L. dr*

**7'. Pterostigmas short and spanning only two cells** **A A**. Thorax green or coppery, with a blue pruinescence only on sides, and only in ♂ **B B**. Cerci of ♂ long and extending beyond middle of cercoids **C C**.
................................. **8**

**8.** ♂ Cerci almost straight, with a spatulate tip **C**, not curved or converging at tip. ♀ Ovipositor not extending beyond end of terminal tubercle or of cercoids **D**.
......................*Lestes sponsa*

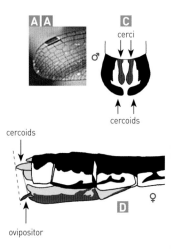

**NORTH:** JFMAM**JJASO**ND
**SOUTH:** JFMAM**JJASO**ND

**DISTRIBUTION:** generally common throughout temperate parts of our region.

**HABITAT:** stagnant, acid or alkaline, fresh or brackish permanent or temporary waters; up to 2,500m.

**CONFUSION SPECIES:** *L. dryas* (in particular) and *L. macrostigma*.

**8'.** ♂ Cerci long, curved and converging at tip **C**. ♀ Ovipositor extending beyond end of terminal tubercle or of cercoids **D**.
......................*Lestes dryas*

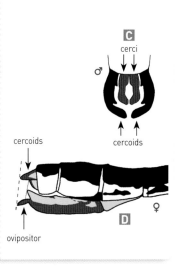

**NORTH:** JFMA**MJJASO**ND
**SOUTH:** JFMA**MJJASO**ND

**DISTRIBUTION:** similar to that of *L. sponsa*, but rarer and more localised relatively common in south of range.

**HABITAT:** all kinds of stagnant waterbodies, occasionally in open woodlands; up to >2,000m.

**CONFUSION SPECIES:** *L. sponsa* (particularly ♀) and *L. macrostigma*.

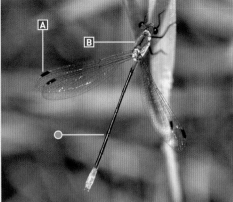

● Body robust, metallic green, with marked blue pulverulence on the back, on the sides of the thorax, as well as at the tip of the abdomen, in both sexes

♂ abdomen: 31–38mm

♀ abdomen: 31–36mm

## *Lestes sponsa* | Common Spreadwing/Emerald Damselfly

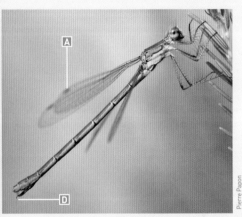

Pierre Papon

♂ abdomen: 25–33mm

♀ abdomen: 25–30mm

## *Lestes dryas* | Robust Spreadwing/Scarce Emerald Damselfly

♂ abdomen: 26–36mm

♀ abdomen: 26–33mm

# Family Platycnemididae

## Genus *PLATYCNEMIS*

**1.** Tibiae of mid- and hindlegs slightly
dilated in ♂, not dilated in ♀, with a
black line extending three-quarters
of their length **A**.
♂ Cercoids strongly bifid at tip **B**.
**Abdomen distinctly orange at maturity**,
with longitudinal black lines only on
segments 7–9 or 7–10 **C**.
♀ Posterior edge of prothorax with a strong
ascending tooth on each side **D**.
.......... ***Platycnemis acutipennis***

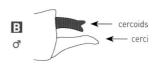

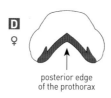

posterior edge
of the prothorax

**1'.** ♂ Cercoids non-bifid, at most simply
notched at tip. Abdomen white or blue
at maturity.
♀ Posterior edge of prothorax at most with
a small median, forward-projecting tooth.
Tibiae of mid- and hindlegs dilated.
.................................. **2**

**2.** Tibiae of mid- and hindlegs moderately
dilated in ♂, more weakly dilated in ♀,
marked with a longitudinal black line along
their entire length at maturity **A**.
♂ Cercoids distinctly notched at tip **B**.
**Abdomen white in immature stages, then
light blue with longitudinal dorsal black
lines along all segments at maturity C**.
♀ Posterior edge of prothorax lacks lateral
teeth **D**.
.......... ***Platycnemis pennipes***

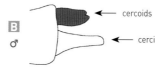

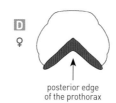

posterior edge
of the prothorax

**2'.** Tibiae of the mid- and hindlegs strongly
dilated in ♂ and moderately dilated in ♀,
the hindlegs at least without a complete
black line **A**.
♂ Cercoids barely notched at tip **B**.
**Abdomen entirely white and always
lacking dorsal black lines on segments
1–5 C**.
♀ Posterior edge of prothorax with a small,
forwards-facing median tooth **D**.
.............. ***Platycnemis latipes***

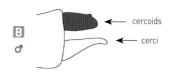

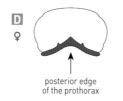

posterior edge
of the prothorax

**NORTH AND SOUTH:** JFMAMJJASOND

**DISTRIBUTION:** Iberian Peninsula and
south and west half of France.

**HABITAT:** running and stagnant waters
up to 1,150m.

**CONFUSION SPECIES:** the immature ♀
of our three *Platycnemis* species are
difficult to distinguish from one another

**NORTH:** JFMAMJJASOND
**SOUTH:** JFMAMJJASOND

**DISTRIBUTION:** very common across most
of Europe, but absent from Ireland, the
far north and most of Iberia.

**HABITAT:** sunny, running and stagnant
neutral or alkaline waters. Recorded
to 1,800m.

**CONFUSION SPECIES:** for immature ♂,
with *P. latipes*; for immature ♀, with
*P. acutipennis* or *P. latipes*.

**NORTH AND SOUTH:** JFMAMJJASOND

**DISTRIBUTION:** Iberian Peninsula and
south-western half of France.

**HABITAT:** flowing waters, especially cal
bends of streams and both small and
large rivers; below 900m.

**CONFUSION SPECIES:** for ♂, with
immature *P. pennipes*.

## *Platycnemis acutipennis* | Orange Featherleg/Orange White-legged Damselfly

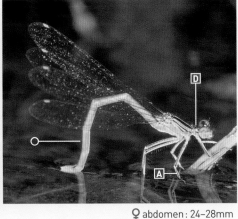

● In most cases, two pale lines on the head (one at the front, one at the back) are present in the three *Platycnemis* species.

⚠ In melanic forms, the line at the back of the head is absent

● Abdomen orange-brown with fine black lines in females

♂ abdomen: 25–30mm

♀ abdomen: 24–28mm

## *Platycnemis pennipes* | Blue Featherleg/White-legged Damselfly

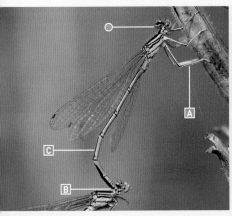

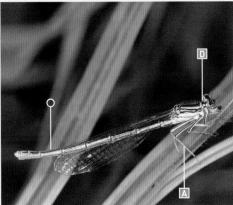

● In most cases, two pale lines on the head (one at the front, one at the back) are present in the three *Platycnemis* species

● Body blue to more or less brownish-pale green, with black markings at maturity (♀)

♂ abdomen: 26–32mm

♀ abdomen: 27–33mm

## *Platycnemis latipes* | White Featherleg

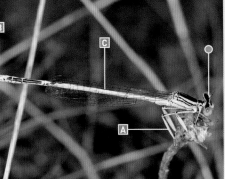

● In most cases, two pale lines on the head (one at the front, one at the back) are present in the three *Platycnemis* species

● ♀ Pale brownish colour with or without black lines

♂ abdomen: 25–30mm

♀ abdomen: 26–30mm

# Family Coenagrionidae

**1. Pterostigmas bicoloured, especially in ♂** 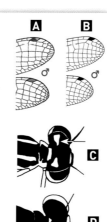.
Abdominal segments 3–7 with black dorsal coloration
in both sexes. ♀ with a weak vulvar spine.
. . . . . . . . . . . . . . . . . . . . . . . . . . . . . . . . . . *Ischnura* (p. 54)

**1'. Pterostigmas uniformly coloured** B. Abdomen of
♂ with a dorsal coloration that is either blue, red
or greenish with black patterns, or metallic green.
♀ with a strong vulvar spine or lacking a vulvar
spine. . . . . . . . . . . . . . . . . . . . . . . . . . . . . . . **2**

**2.** Occiput marked with clear postocular spots, which
may darken with age C . . . . . . . . . . . . . . . . . . . . . . **3**

**2'.** Occiput lacking postocular spots, uniformly brown
or black D . . . . . . . . . . . . . . . . . . . . . . . . . . . . . . **6**

**3.** A very small species, abdomen ≤23mm, hindwings
≤16mm. **Upperside of abdomen and of thorax
metallic green to coppery** A . Thorax lacks clear
antehumeral bands B . . . . . . *Nehalennia speciosa*

**3'.** A larger species, body blue or green with black
markings. Abdomen and thorax lack metallic
reflections; thorax always with two clear
antehumeral bands. . . . . . . . . . . . . . . . . . . . . . . **4**

**4. Median lateral sutures of the thorax without
black** A . Second abdominal segment of ♂ with
a black mushroom-shaped dorsal mark B .
♀ with a strong vulvar spine C .
. . . . . . . . . . . . . . . . . . . . . . *Enallagma cyathigerum*

**4'. Median lateral sutures of thorax at least partly
lined with black** E . Second abdominal segment
of ♂ with a black dorsal mark not in the shape of a
mushroom. ♀ lacks vulvar spine . . . . . . . . . . . . . . . **5**

**5.** ♂ Cercoids pincer-shaped, longer than the 10th
abdominal segment F . Second abdominal segment
with a black marking along its entire length.
♀ Posterior edge of prothorax slightly sinuous.
Pale areas of thorax, sides of segments 1 (2) and
8–10 and cercoids mostly light yellow, but sides
of segments 3–7 blue G .
. . . . . . . . . . . . . . . . . *Erythromma* (ex *Cercion*) (p. 56)

**5'.** Species not as above. Cercoids of ♂ equalling the
10th abdominal segment or shorter H .
. . . . . . . . . . . . . . . . . . . . . . . . . . . . *Coenagrion* (p. 48)

**6. General colour black and blue, or black and green.**
Cercoids of ♂ much longer than cerci.
. . . . . . . . . . . . . . . . . . . . . . . *Erythromma* (p. 56)

**6'. General colour red, or orange and black** (but some
♀ are entirely black). Cercoids of ♂ slightly longer
or shorter than cerci . . . . . . . . . . . . . . . . . . . . . . **7**

**NORTH :** JFMAMJJASOND
**SOUTH :** JFMAMJJASOND

**DISTRIBUTION:** highly threatened in
most western European countries,
due to the gradual disappearance
of its biotopes and global warming.
Particularly threatened in Switzerland.
Recently found in France and the
Netherlands. Extinct in Belgium and
Luxembourg. Absent from eastern,
southern and northern areas.

**HABITAT:** stagnant oligotrophic and
mesotrophic waters; mainly inhabits
marshy banks of *Sphagnum* peat
bogs and mesotrophic marshes with
bulrushes, sedges, horsetails, rushes
and Marsh Cinquefoils; up to 900m.

**CONFUSION SPECIES:** at maturity, can
superficially be confused with *Ischnura
pumilio*. Specimens with intermediate
coloration can resemble small *Lestes*
species.

**NORTH :** JFMAMJJASOND
**SOUTH :** JFMAMJJASOND

**DISTRIBUTION:** common across the
region, except in Corsica.

**HABITAT:** all kinds of stagnant waters;
from sea level to more than 2,500m.

**CONFUSION SPECIES:** for ♀, with those of
*Coenagrion lunulatum*.

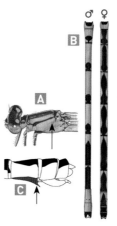

10th s. | cercoids

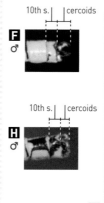

10th s. | cercoids

## *Nehalennia speciosa* | Sedgling/Pygmy Damselfly

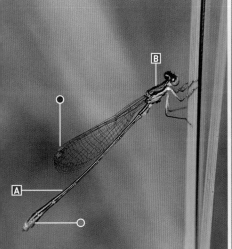

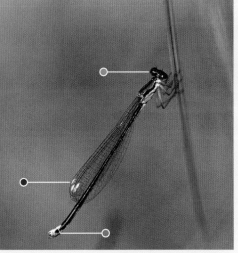

● A pale line on the head

● Pterostigmas grey-whitish to brown

● Tip of the abdomen and sides of the thorax blue at maturity in males

♂ abdomen: 19–23mm

♀ abdomen: 19–22mm

## *Enallagma cyathigerum* | Common Bluet/Common Blue Damselfly

● Pale antehumeral bands almost as wide as the humeral band

♂ abdomen: 22–30mm

♀ abdomen: 22–29mm

## Genera *PYRRHOSOMA* and *CERIAGRION*

**7. Legs black** . Thorax with red to yellow antehumeral bands **B**. **Pterostigmas black C**.
Cercoids and cerci of ♂ distinctly as long as 10th abdominal segment.
............. *Pyrrhosoma nymphula*

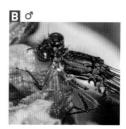

NORTH: JFMAMJJASOND
SOUTH: JFMAMJJASOND

DISTRIBUTION: very common across the region.

HABITAT: stagnant or weakly flowing water, including brackish waters (pools, ponds, lakes, peat bogs, ditches, coastal lagoons, slow-flowing rivers); >2,000m.

CONFUSION SPECIES: *Ceriagrion tenellum*

**7'. Legs reddish or yellowish** **A**.
Thorax without antehumeral bands, usually with a black dorsal coloration **B**.
**Pterostigmas red in ♂, yellowish brown to more or less grey in ♀ C**. Cercoids shorter than cerci and not exceeding a third of length of 10th segment.
............... *Ceriagrion tenellum*

NORTH: JFMAMJJASOND
SOUTH: JFMAMJJASOND

DISTRIBUTION: more frequent and widespread in the south and south-west of the region. Absent from Fennoscandia and eastern Europe.

HABITAT: stagnant or weakly flowing calcareous, neutral or very acidic waters, even in peat bogs; up to 1,000m.

CONFUSION SPECIES: *Pyrrhosoma nymphula*.

**Note on *Ceriagrion tenellum*:** andromorphic (*erythrogastrum*) females, which are a minority morph, have an entirely red abdomen. In heteromorphic females the abdomen has a black or red and black dorsal coloration. These include a morph called *typica*, forming the majority, in which the dorsal surface of segments 4(or 5)–8 is black; a morph *intermedium*, in which the dorsal surface of the 7th segment and the apical part of segments 3–6 are marked black; and a third morph called *melanogastrum*, in which the dorsal surface of the abdomen is completely black.

## *Pyrrhosoma nymphula* | Large Red Damselfly

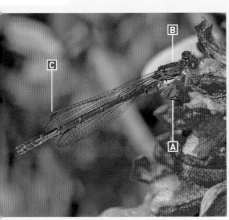

♂ abdomen: 24–31mm

♀ abdomen: 25–30mm

## *Ceriagrion tenellum* | Small Red Damselfly

♂ abdomen: 22–28mm

♀ abdomen: 23–29mm

## Genus *COENAGRION*

### IDENTIFICATION OF MALES

**1.** ♂ Cercoids longer than cerci 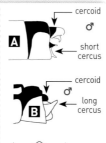A .................. **2**

**1'.** ♂ Cercoids shorter than cerci B ................ **5**

**2.** ♂ Black dorsal marking of second abdominal segment not in a 'U' or 'V' shape......................... **3**

**2'.** ♂ Black dorsal marking of second abdominal segment in the shape of a more or less well defined 'U' or 'V', attached or not to tip of segment ................. **4**

### IDENTIFICATION OF FEMALES

**a.** ♀ Posterior edge of prothorax with a strong median incision C.
............... ***Coenagrion caerulescens*** (p. 50)

**a'.** ♀ Posterior edge of prothorax not as above ........ **b**

**b.** ♀ Posterior edge of prothorax not trilobed, but simply bearing a median tubercle....................... **c**

**b'.** ♀ Posterior edge of prothorax more or less trilobed. .... **e**

**3.** ♂ **Black dorsal marking of second abdominal segment formed of three separate lunules** A. Black markings of segments 3–6 covering much more than half of their length B. Cercoids thick, short, not converging at tip C. Posterior edge of prothorax trilobed D.
.......................... ***Coenagrion lunulatum***

**c.** ♀ Eighth abdominal segment bicoloured dorsally, with black pattern covering more than half of its length and ending anteriorly in a triangle shape, revealing clear margins. Rest of abdomen with a black dorsal coloration B. **Posterior edge of prothorax with a strongly marked median tubercle**
.......................... ***Coenagrion lunulatum***

**c'.** ♀ Eighth abdominal segment entirely black. Posterior edge of prothorax with a moderately marked median tubercle E. ................................... **d**

**3'.** ♂ **Black dorsal marking of second abdominal segment in shape of a bull's head**, highly variable A. Markings of segments 3–6 covering approximately half their length B. Cercoids elongated and converging at tip C. Posterior edge of prothorax not trilobed D.
....................... ***Coenagrion mercuriale***

**d.** ♀ Posterior edge of prothorax almost linear, with a small median lobe E. ..... ***Coenagrion mercuriale***

**d'.** ♀ Posterior edge of prothorax with oblique sides on either side of the median lobe.
.................. ***Coenagrion hastulatum*** (p. 50)

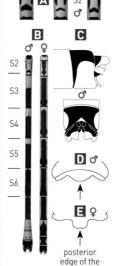
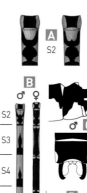
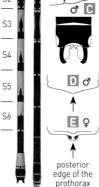

cercoid
♂
short cercus

cercoid
♂
long cercus

posterior edge of the prothorax ♀

S2 ♂

S2
S3
S4
S5
S6

♂ ♀
♂
♂
D ♂
E ♀
posterior edge of the prothorax

S2

S2
S3
S4
S5
S6

♂ ♀
♂ C
D ♂
E ♀
posterior edge of the prothorax

NORTH: JFM**AMJJ**ASOND
SOUTH: JFM**AMJJ**ASOND

**DISTRIBUTION:** northern Europe, plus parts of Ireland and France. Has disappeared from Switzerland (last observation in 1989), and scarce and local in Belgium and the Netherlands. Absent from Britain.

**HABITAT:** stagnant mesotrophic and oligotrophic waters, including peat lakes and ponds, *Sphagnum* peat bog and marshes with sedges, bulrushes, cottongrasses, horsetails or moorgrasses; up to 1,250m.

**CONFUSION SPECIES:** for ♂, with those C. hastulatum bearing lunules. For ♀, with those of *Enallagma cyathigerum*.

NORTH: JFM**AMJJAS**OND
SOUTH: JFM**AMJJAS**OND

**DISTRIBUTION:** Locally common in south-west Britain, France and Iberian Peninsula; does not occur east of Austria, Germany and Italy.

**HABITAT:** high-quality, sunny, moderately-flowing alkaline waters; rarely in acidic or brackish waters (springs, brooks, meadow streams, ditches and channels with aquatic and hygrophilic vegetation); generally below 700m.

**CONFUSION SPECIES:** for ♂, with those C. scitulum and C. caerulescens.

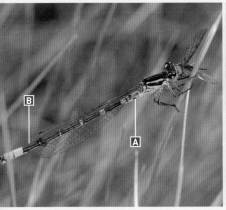

♂ abdomen: 23–36mm

♀ abdomen: 22–26mm

*Coenagrion mercuriale* | Mercury Bluet/Southern Damselfly

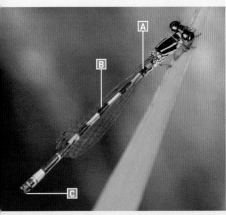

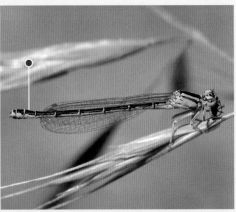

● Mature females generally greenish, dorsal side of the abdomen entirely green, including on the 8th segment; rare andromorphic morphs

♂ abdomen: 19–27mm

♀ abdomen: 21–27mm

**4.** ♂ **Cercoids equal to half the length of 10th abdominal segment**, and curved inwards in dorsal view 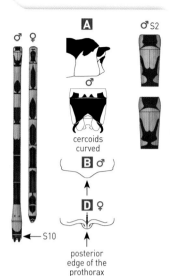**A**.
Posterior edge of prothorax with a large median lobe **B**.

**e.** ♀ Pterostigmas approximately twice as long as wide **C**. Posterior edge of prothorax moderately trilobed, **median lobe not incised D**.
. . . . . . . . . . . . . . . *Coenagrion scitulum*

**e'.** ♀ Pterostigmas less than twice as long as wide. Posterior edge of prothorax variable.
. . . . . . . . . . . . . . . . . . . . . . . . . . . . . . . . **f**

cercoids curved

posterior edge of the prothorax

NORTH: JFMAMJJASOND
SOUTH: JFMAMJJASOND

DISTRIBUTION: widespread in much of the region. Range is expanding northwards.

HABITAT: sunny, stagnant, non-brackish waters, colonised by hydrophytes.

CONFUSION SPECIES: separating this species from *C. caerulescens* in the field is difficult.

**4'.** ♂ **Cercoids longer than half 10th abdominal segment**, and straight in dorsal view **A**. Posterior edge of prothorax with a very small median lobe **B**.
. . . . . . . . . *Coenagrion caerulescens*

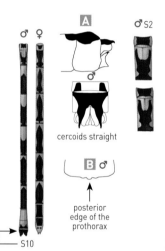

cercoids straight

posterior edge of the prothorax

NORTH AND SOUTH: JFMAMJJASOND

DISTRIBUTION: around the western Mediterranean, where it remains rare overall.

HABITAT: sunny, shallow, high-quality flowing waters, with vegetation such as water milfoils; up to 1,100m.

CONFUSION SPECIES: *C. scitulum*.

**5.** ♂ **Posterior edge of prothorax not trilobed C**. Cercoids very short **B**. Black dorsal marking of second abdominal segment in shape of a halberd or ace of spades, surrounded laterally by two lines; this marking sometimes reduced to three separate lunules **A**.
. . . . . . . . . . . . .*Coenagrion hastulatum*

**5'.** ♂ Species not as above. Posterior edge of prothorax more or less trilobed.
. . . . . . . . . . . . . . . . . . . . . . . . . . . . . . . **6**

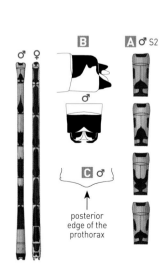

posterior edge of the prothorax

NORTH: JFMAMJJASOND
SOUTH: JFMAMJJASOND

DISTRIBUTION: northern and eastern parts of the region. In the Scottish Highlands, restricted to a few lochans.

HABITAT: stagnant acidic waters (*Sphagnum* peat bogs, marshes with sedges and bogbeans); up to 2,500m.

CONFUSION SPECIES: for ♂, with those of *C. ornatum* and *C. lunulatum*.

## *Coenagrion scitulum* | Dainty Bluet/Dainty Damselfly

● 6th abdominal segment almost entirely black

● Black pattern thick and U-shaped, attached to the tip of the second segment

♂ abdomen: 20–27mm

♀ abdomen: 21–26mm

## *Coenagrion caerulescens* | Mediterranean Bluet

● The identification of males is difficult in the field because the black markings are highly variable

♂ abdomen: 18–26mm

♀ abdomen: 19–27mm

## *Coenagrion hastulatum* | Spearhead Bluet/Northern Damselfly

♂ abdomen: 22–34mm

♀ abdomen: 24–32mm

**6.** ♂ **Black dorsal marking of abdominal
segments 3–4(5) in shape of a spearhead
A**. Marking of second abdominal segment
in shape of a halberd surrounded by two
lateral lines, these generally fused with it **B**.
............... *Coenagrion ornatum*

**6'.** ♂ Black dorsal marking of abdominal
segments 3–5 different; marking of second
abdominal segment either in the shape of
a U separated from the distal end of the
segment, or in the shape of a Y fused with
the tip of the segment.
.................................. **7**

**f.** ♀ **Median lobe of posterior edge of prothorax bilobed C**. Abdominal segments 3–8
bicoloured dorsally, with black markings on
segments 3 and 4 elongated into a point or
tridentate **D**.
............... *Coenagrion ornatum*

**f'.** ♀ Median lobe of prothorax not bilobed.
Dorsal aspect of abdomen usually mostly
black, rarely blue and black.
.................................. **g**

**7.** ♂ Second abdominal segment with a thick
Y-shaped black dorsal marking, attached
to the end of the segment (rarely U-shaped
and detached from end of segment) **A**.
**Posterior edge of prothorax strongly
trilobed C**. Upper branch of cerci slightly
longer than cercoids or of same length **B**.
Black markings highly variable.
............ *Coenagrion pulchellum*

**g.** ♀ **Posterior edge of the prothorax
strongly trilobed D**. Highly variable black
markings.
............ *Coenagrion pulchellum*

**g'.** ♀ **Posterior edge of prothorax weakly
trilobed D**.
............... *Coenagrion puella*

**7'.** ♂ Black dorsal marking of second
abdominal segment in the shape of an
isolated U (rarely fragmented or attached
to end of segment) **A**. **Posterior edge
of prothorax weakly trilobed C**. Upper
branch of cerci 1.5–2 times longer than
cercoids **B**.
............... *Coenagrion puella*

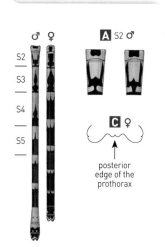

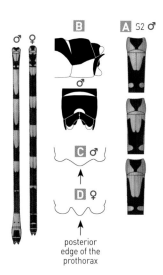

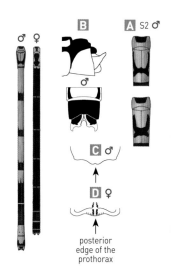

**DISTRIBUTION:** common in parts of Fra▮
and Switzerland. Disappeared from
north-east France; nearly extinct in Ita▮

**HABITAT:** silted streams, ditches and
hillside seepages in sunny meadows.

**CONFUSION SPECIES:** *C. hastulatum*. For
♀, with certain andromorphic forms
of *C. puella*.

**DISTRIBUTION:** locally abundant across
much of the region; uncommon in Bri▮

**HABITAT:** sunny, stagnant mesotrophic
and eutrophic fresh waters with aquat▮
vegetation; up to 1,200m.

**CONFUSION SPECIES:** for ♂ and ♀, with
*C. scitulum* and *C. puella*; for ♀, with
*C. mercuriale* and *C. hastulatum*.

**DISTRIBUTION:** abundant across the
region and one of our most common
odonates.

**HABITAT:** permanent and stagnant or
slow-flowing fresh waters; up to 2,200m▮

**CONFUSION SPECIES:** for ♂ and ♀,
with *C. pulchellum*; for ♀, also with
*C. mercuriale* and *C. hastulatum*.

## *Coenagrion ornatum* | Ornate Bluet

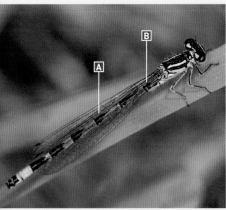

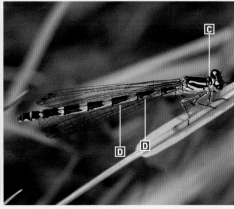

♂ abdomen: 20–30mm

♀ abdomen: 23–29mm

## *Coenagrion pulchellum* | Variable Bluet/Variable Damselfly

● Fine lateral black lines

♂ abdomen: 24–32mm

♀ abdomen: 23–31mm

## *Coenagrion puella* | Azure Bluet/Azure Damselfly

● Black markings on segments 3–5 are usually short in dorsal view

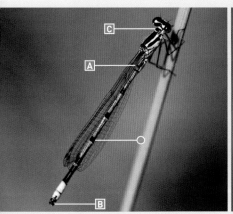

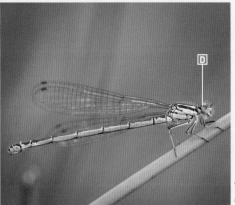

Pierre Papon

♂ abdomen: 22–32mm

♀ abdomen: 23–30mm

## Genus *ISCHNURA*

1. Pterostigmas of forewings larger than those of hindwings, especially in ♂ **A**. Posterior edge of prothorax rounded. ♂**Ninth abdominal segment dorsally blue or blue with variable black spots;** eighth abdominal segment black or bicoloured **B**. Cercoids simple, without a differentiated internal branch **C**. ♀Eighth abdominal segment dorsally black **D**.
   . . . . . . . . . . . . . . . . . . *Ischnura pumilio*

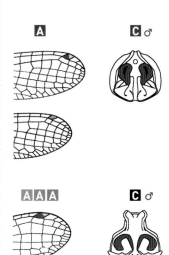

NORTH: JFMAMJJASOND
SOUTH: JFMAMJJASOND

DISTRIBUTION: abundant in recently crea habitats, but less common in the north

HABITAT: new water features.

CONFUSION SPECIES: other *Ischnura* speci

1'. Pterostigmas almost identical on the four wings **A A A**. Posterior edge of prothorax with a differentiated median tubercle. ♂ **Eighth abdominal segment dorsally blue**; ninth abdominal segment black **B B B**. Cercoids with a differentiated internal branch, arched or curved **C**.
   ♀ Eighth abdominal segment mostly clear dorsally . . . . . . . . . . . . . . . . . . . . . . . . . . **2**

NORTH AND SOUTH : JFMAMJJASOND

DISTRIBUTION: south-western France.

HABITAT: stagnant or weakly flowing waters; sometimes in brackish waters

CONFUSION SPECIES: *I. elegans*.

2. Internal branches of ♂ cercoids initially convergent, then divergent **E**. **In dorsal view, posterior edge of prothorax differentiated into a short strap D**.
   . . . . . . . . . . . . . . . . . *Ischnura graellsii*

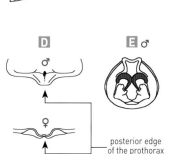

posterior edge of the prothorax

2'. Internal branches of ♂ cercoids parallel, convergent or crossing, never divergent in their second half . . . . . . . . . . . . . . . . . **3**

NORTH: JFMAMJJASOND
SOUTH: JFMAMJJASOND

DISTRIBUTION: most of the region.

HABITAT: sunny, stagnant waters.

CONFUSION SPECIES: *I. graellsii*.

3. **Posterior lobe of prothorax rising in both sexes, forming a large median strap clearly visible** in dorsal or posterior view **D**. Internal branches of ♂ cercoids not crossing **E**.
   . . . . . . . . . . . . . . . . . *Ischnura elegans*

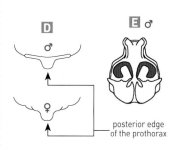

posterior edge of the prothorax

3'. Posterior edge of prothorax composed of several lobes (sometimes not very distinct), generally appearing bilobed in posterior view, with the addition of a median tubercle, this especially visible in dorsal view in ♀ **D**. Internal branches of ♂ cercoids crossing **E**.
   . . . . . . . . . . . . . . . . . . *Ischnura genei*

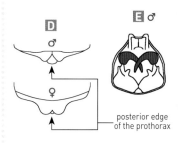

posterior edge of the prothorax

NORTH AND SOUTH: JFMAMJJASOND

DISTRIBUTION: endemic to the islands of the Tyrrhenian Sea.

HABITAT: sunny, stagnant waters; to 1,000m.

CONFUSION SPECIES: *I. elegans*.

### *Ischnura pumilio* | Scarce Blue-tailed Damselfly

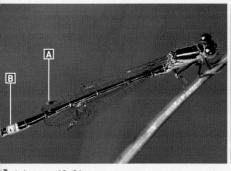

♂ abdomen : 19–26mm

♀ abdomen : 18–27mm

### *Ischnura graellsii* | Iberian Bluetail

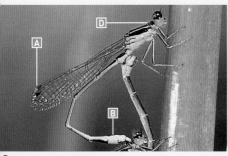
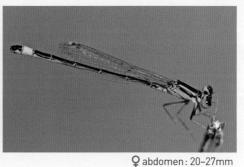

♂ abdomen : 19–25mm

♀ abdomen : 20–27mm

### *Ischnura elegans* | Common Bluetail/Blue-tailed Damselfly

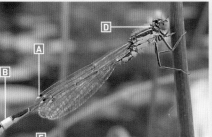

♂ abdomen : 22–29mm

♀ abdomen : 22–29mm

### *Ischnura genei* | Island Bluetail

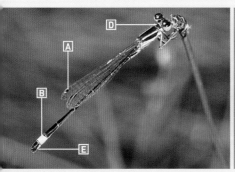
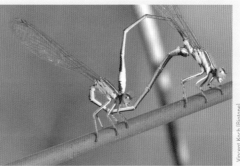

Vincent Koch |Biotopel

♂ abdomen : 18–26mm

♀ abdomen : 18–27mm

## Genus *ERYTHROMMA*

**1.** ♂ **Cercoids longer than 10th abdominal segment** . Marking of second abdominal segment covering its entire length **B**. Dorsal surface of abdomen black and blue. Eyes blue **C**.
♀ **Posterior edge of the prothorax almost straight D**. Sides of abdomen yellowish on segments 1(2) and 8–10; blue on the other segments **E**. Cercoids yellowish.
. . . . . . . . . . . . . . . *Erythromma lindenii*

**1'.** ♂ **Cercoids the same length as 10th abdominal segment.**
Eyes red **C C**.
Dorsal surface of abdomen entirely black on segments 2–8 **D D**.
♀ **Posterior edge of prothorax not straight.** Sides of abdomen blue. Cercoids black.
. . . . . . . . . . . . . . . . . . . . . . . . . . . **2**

**2.** ♂ **Thorax lacking clear antehumeral stripes A**. **10th abdominal segment blue B**. Cercoids enlarged in their second half.
♀ **Tarsi black F** and posterior edge of prothorax trilobed in dorsal view **E**.
. . . . . . . . . . . . . . . .*Erythromma najas*

**2'.** ♂ **Thorax with antehumeral stripes A**. **10th abdominal segment blue** with an X-shaped black dorsal mark extending on to cercoids **B**. Cercoids narrowing in their second half.
♀ **Tarsi marked dorsally with a yellow band F**. Posterior edge of prothorax rounded in dorsal view **E**.
. . . . . . . . . . . *Erythromma viridulum*

chalice-shaped pattern

**B**

♂ ♀

**B** S2 ♂

S2
S3
S4
S5
S6
S7
S8
S9

**D** ♀

posterior edge of the prothorax

**A**

cercoids longer than the 10th segment

**E** ♀

posterior edge of the prothorax

**E** ♀

posterior edge of the prothorax

**NORTH :** JFMAM**JJAS**OND
**SOUTH :** JFMAM**JJAS**OND

**DISTRIBUTION:** common in the south, gradually expanding northward in all regions; absent from Britain, Ireland, Scandinavia and north-east Europe.

**HABITAT:** sunny, flowing or stagnant waters; up to 1,100m.

**CONFUSION SPECIES:** ♀ resembles those of many *Coenagrion*.

**NORTH :** JFMAM**JJAS**OND
**SOUTH :** JFMAM**JJAS**OND

**DISTRIBUTION:** becoming rare in the Mediterranean region and only local in the north of its range. Absent from Ireland, the Iberian Peninsula and Greece.

**HABITAT:** sunny, stagnant or weakly flowing waters containing floating plants; up to 1,300m.

**CONFUSION SPECIES:** ♀ resembles those of many *Coenagrion*.

**NORTH :** JFMAM**JJAS**OND
**SOUTH :** JFMAM**JJAS**OND

**DISTRIBUTION:** common in the south and expanding in the northern part of its range; first recorded in Britain in 1999.

**HABITAT:** sunny, stagnant or weakly flowing waters with abundant surface vegetation; up to 800m.

**CONFUSION SPECIES:** for ♀, with *E. najas*, *Coenagrion* and *Ischnura*.

## *Erythromma lindenii* | Blue-eye/Goblet-marked Damselfly

● Black spear-shaped marks on segments 3–6

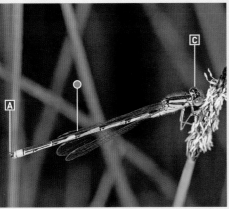

♂ abdomen: 21–30mm

♀ abdomen: 21–29mm

## *Erythromma najas* | Large Redeye/Red-eyed Damselfly

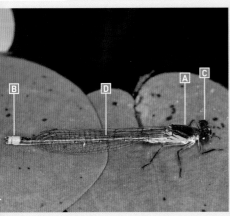

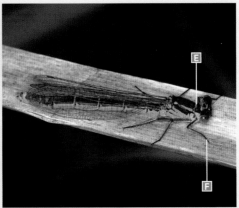

♂ abdomen: 25–31mm

♀ abdomen: 26–30mm

## *Erythromma viridulum* | Small Redeye/Small Red-eyed Damselfly

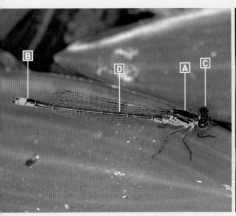

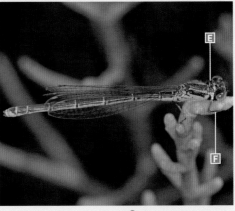

♂ abdomen: 22–29mm

♀ abdomen: 22–28mm

# Family Aeshnidae

**1.** Transverse veins present in medial area (ma), between base of wing and arculus **A**.
.............................**Boyeria irene**

**1'.** No transverse veins in medial area (ma), between the base of wing and the arculus **A**.
......................................... **2**

**2.** Ra vein inserted on arculus, closer to R+M than to Cu. R3 with an abrupt sinuosity under pterostigma **B**.
♂ Base of hindwings without anal triangle or anal angle.
♂ No lateral auricles on second abdominal segment .................................. **3**

**2'.** Ra vein inserted on the arculus in its centre, or closer to Cu than to R+M. R3 with a wide and regular curve under pterostigma **C**.
♂ Base of hindwings with an anal triangle (at).
♂ Lateral auricles present on second abdominal segment **D**. .......................... **4**

**3.** Two (very rarely three) cells present beyond discoidal cell of hindwings, followed by a row of cells between anal and cubital veins **E**. Sides of abdomen with distinct longitudinal lateral carinae.
♂ Cercoids truncated at tip or with a point or external lateral spine **F**.
.............................. **Anax** (p. 66)

**3'.** Three cells present beyond discoidal cell of hindwings, followed by a row of cells between anal and cubital veins **A**. Abdomen brownish, without longitudinal lateral carinae.
♂ Cercoids with a median apical tip **B**. Sides of thorax green **C**.
..................**Hemianax ephippiger**

**4.** One or two rows of cells between IR3 and Srv veins on one hand, and between M and Smv veins on other hand **G**.
♂ Hindwings with a weakly marked anal angle (aa).
..............**Brachytron pratense** (p. 64)

**4'.** More than two rows of cells between IR3 and Srv veins on one hand, and between M and Smv veins on the other hand.
♂ Hindwings with a strongly marked anal angle (aa) **H**.
.......................... **Aeshna** (p. 60)

## Boyeria irene | Western Spectre/Dusk Hawker

- Green eyes touching along a line
- 'Camouflage-like' pattern, with brown, green-grey to blue-grey markings
- Abdomen narrowed at base
- Segments 9 and 10 pale (♂)

**NORTH:** JFMAM**JJASO**ND
**SOUTH:** JFMAM**JJAS**OND

**DISTRIBUTION:** common in the western Mediterranean area and western France. The species also occurs in central Switzerland.

**HABITAT:** flowing waters, preferably shaded; sometimes also in lakes; up to 1,000m in the southern Alps and 1,300m on Corsica.

**CONFUSION SPECIES:** its flight style resembles that of *Brachytron pratense*, but *Boyeria irene* inhabits rivers, while *B. pratense* prefers ponds. From a distance, *B. irene* can also be confused with a *Cordulegaster*.

**NOTE:** some females have short cercoids (*brachycerca* morph); others have cercoids as long as those of males. The proportion of the two morphs is quite variable, and those with short cercoids can represent 40–100 per cent of females.

♂ abdomen : 53–56mm

♀ abdomen : 49–54mm

## Hemianax ephippiger | Vagrant Emperor

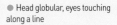

- Head globular, eyes touching along a line
- ♂ Blue marking only visible on S2 (unlike in *A. parthenope*)
- ♀ No marking, or a greyish marking limited to the dorsal side of the second abdominal segment

**NORTH AND SOUTH:** JFM**AMJJASO**ND

**DISTRIBUTION:** a migratory species, breeding in the Mediterranean region and in some years occasionally reaching as far north as Iceland.

**HABITAT:** stagnant waters, even those that are temporary or brackish, including coastal marshes, rice fields, permanent or temporary pools, ponds, dam lakes and gravel pits.

**CONFUSION SPECIES:** *Anax parthenope*.

♂ abdomen : 46–56mm

♀ abdomen : 44–50mm

## Genus *AESHNA*

**1.** General colour of body brown to red, with a small yellow dorsal marking in the shape of an isosceles triangle on second abdominal segment; body without blue or green spots **A**.
Base of the hindwings saffron yellow **B**.
♂ Anal triangle of hindwings consisting of (3)4–6 cells.
................... *Aeshna isoceles*

**1'.** Abdomen either brown or black with conspicuous blue, green or yellow spots; or reddish brown with small blue and/or yellow spots. Wings either translucent or smoky.
♂ Anal triangle of hindwings consisting of two or three cells.
.................................. **2**

**2.** Incomplete T-shaped black spot present on frons **A**.
General colour brown, with blue spots in ♂ and/or very small yellowish spots in ♀ **B**.
Wings entirely smoky, with reddish-brown veins **C**.
.................. *Aeshna grandis*

**2'.** T-shaped black spot present on frons **A**.
General colour dark brown and blue **B**, yellow or green. Wings entirely translucent **C**.
.................................. **3**

**3.** Sides of thorax with two clear lines, these approximately a fifth of the width of brown band separating them **D**.
................... *Aeshna caerulea*

**3'.** Sides of thorax either with two clear lines wider than a fifth of brown band separating them, or almost entirely clear.
.................................. **4**

**4.** Suture separating forehead from post-clypeus highlighted with a clear black line, this a little wider than the suture itself in its middle **A**.
♂ Anal triangle consisting of two cells.
.................................. **5**

**4'.** Suture separating forehead from post-clypeus at most slightly darkened **B**.
♂ Anal triangle consisting of three cells.
.................................. **6**

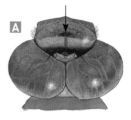

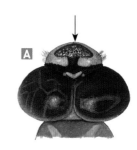

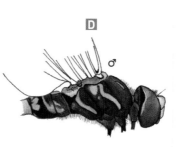

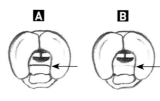

## *Aeshna isoceles* | Green-eyed Hawker/Norfolk Hawker

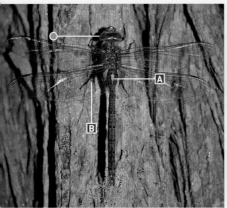

● Eyes large and globular, a striking emerald green at maturity, adjoining over a great length

Thomas Roussel (Biotope)

♂ abdomen: 47–50mm

♀ abdomen: 49–54mm

## *Aeshna grandis* | Brown Hawker

● Two bold yellow bands on the side of the thorax

Thomas Roussel (Biotope)

♂ abdomen: 50–60mm

♀ abdomen: 49–55mm

## *Aeshna caerulea* | Azure Hawker

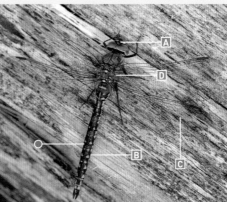

● ♂ Eyes vidid blue, touching along a line; black T-shaped marking on the frons

● ♂ Blue mediodorsal and posterodorsal spots of the same size. ♀ Dorsal spots pale and very small; lateral spots large, converging into an almost continuous yellow band

♂ abdomen: 43–48mm

♀ abdomen: 42–44mm

**5.** ♂ Abdominal segments with pale yellow mediodorsal spots that are much smaller than posterodorsal spots, the latter bright blue at maturity .
♀ Cercoids arranged in two planes, almost perpendicular to each other and with a slightly convex or straight external margin .
. . . . . . . . . . . . . . . . . . . . . .*Aeshna juncea*

NORTH: JFMAMJJASOND
SOUTH: JFMAMJJASOND

DISTRIBUTION: common in northern areas, and in mountain ranges in the south of the region.

HABITAT: stagnant oligotrophic and acidic waters (sedge marshes, peat bogs, peat ponds); 250–2,500m.

CONFUSION SPECIES: *A. mixta* and *A. subarctica*.

**5'.** ♂ Abdominal segments 3 and 4 with pale mediodorsal and posterodorsal spots of approximately the same size and colour: blue, greenish or yellow 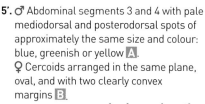.
♀ Cercoids arranged in the same plane, oval, and with two clearly convex margins 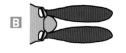.
. . . . . . . . . . . . . . . . .*Aeshna subarctica*

NORTH: JFMAMJJASOND
SOUTH: JFMAMJJASOND

DISTRIBUTION: established in Switzerla and the Vosges mountains; scattered across the north-east of the region. F in region is *elisabethae*.

HABITAT: strictly dependent on *Sphagn* peat bogs; above 1,200m. Coexists wi *A. juncea*, *Somatochlora alpestris* and *S. arctica*.

CONFUSION SPECIES: *A. juncea*.

**6.** Antehumeral and lateral bands wide and green 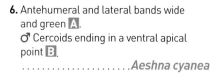.
♂ Cercoids ending in a ventral apical point .
. . . . . . . . . . . . . . . . . . . . . .*Aeshna cyanea*

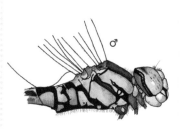

NORTH: JFMAMJJASOND
SOUTH: JFMAMJJASOND

DISTRIBUTION: widespread, but rarely observed in large numbers.

**6'.** Antehumeral bands narrow.
. . . . . . . . . . . . . . . . . . . . . . . . . . . . . . **7**

HABITAT: all stagnant waters, from lowland areas to mountains, even ver acidic *Sphagnum* peat bogs, woodlanc ponds and polluted concrete basins.

CONFUSION SPECIES: for ♀, with *Ophiogomphus cecilia*.

## *Aeshna juncea* | Moorland Hawker/Common Hawker

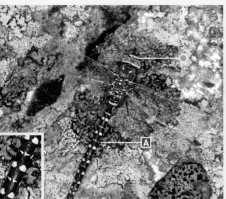

● Black T-shaped spot on the frons

♂ abdomen : 51–59mm

♀ abdomen : 50–57mm

## *Aeshna subarctica* | Bog Hawker/Subarctic Hawker

● Black T-shaped spot on the frons

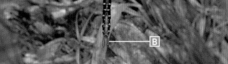

♂ abdomen : 47–57mm

♀ abdomen : 49–55mm

## *Aeshna cyanea* | Blue Hawker/Southern Hawker

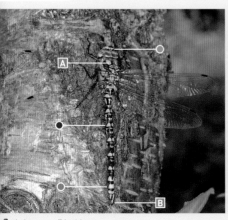

● Black T-shaped spot on the frons

● Irregular green and blue patches on the abdomen

● Fused blue patches on S9 and S10

♂ abdomen : 51–61mm

♀ abdomen : 52–59mm

**7.** Sides of thorax uniform yellow, blue or greenish, with only the sutures finely highlighted in black **A**.
♂ Mediodorsal and posterodorsal blue spots on abdominal segments of the same width **B**. Cercoids with a ventral tooth at base, clearly visible in profile view **C**.
♀ Ovipositor not extending beyond end of ninth abdominal segment **D**.
........................ *Aeshna affinis*

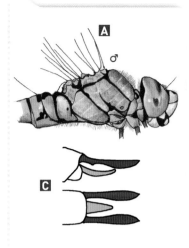

NORTH: JFMAMJJASOND
SOUTH: JFMAMJJASOND

DISTRIBUTION: a locally abundant Mediterranean species, becoming increasingly common in areas with a continental climate.

HABITAT: stagnant waters, even temporary or slightly brackish water Sometimes in calm areas of large rive or in streams with low water levels.

CONFUSION SPECIES: for ♂, with A. caerulea.

**7'.** Sides of thorax with two yellow or yellow-and-blue bands, separated by a broad, well-defined brown band, the three bands being approximately the same width **A**.
♂ Clear mediodorsal spots of abdominal segments much smaller than postero-dorsal spots **B**. Cercoids without ventral teeth at base **C**.
♀ Ovipositor reaching middle of 10th abdominal segment **D**.
........................ *Aeshna mixta*

NORTH AND SOUTH: JFMAMJJASOND

DISTRIBUTION: generally very common across most of Europe, except Scotland and northern Scandinavia.

HABITAT: stagnant mesotrophic or eutrophic waters, even sometimes in brackish waters such as marshes an ponds with open banks.

CONFUSION SPECIES: an A. juncea lookalike. Confusion is also possible with A. affinis.

## Genus *BRACHYTRON*

One or two rows of cells between IR3 and Srv veins on one hand, and between M and Smv veins on the other hand **A**.
♂ Hindwings with a weakly marked anal angle (aa) **B**.
................ *Brachytron pratense*

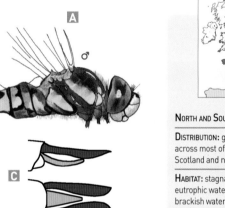

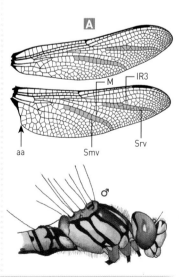

NORTH: JFMAMJJASOND
SOUTH: JFMAMJJASOND

DISTRIBUTION: widespread across mos of the region; less common to the north and in the Iberian Peninsula.

HABITAT: permanent stagnant fresh waters surrounded by beds of sedges reeds or bulrushes; alkaline, neutral or acidic waters; rarely up to 900m.

CONFUSION SPECIES: none.

## Aeshna affinis | Blue-eyed Hawker/Southern Migrant Hawker

- Eyes touching along a line; black T-shaped marking on the frons

Thomas Roussel (Biotope)

♂ abdomen : 39–48mm

♀ abdomen : 42–49mm

## Aeshna mixta | Migrant Hawker

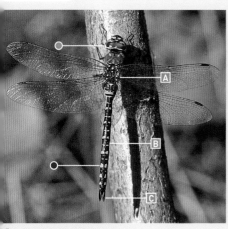

- Eyes touching along a line; black T-shaped marking on the frons

- Pale spots of variable size on the abdomen

♂ abdomen : 44–54mm

♀ abdomen : 43–51mm

## Brachytron pratense | Hairy Hawker/Hairy Dragonfly

- Eyes touching along a very short distance; black T-shaped marking on the frons

- Elongated blue markings

- Abdomen not constricted at the 3rd segment

- Extremely hairy body

♂ abdomen : 41–46mm

♀ abdomen : 38–42mm

## Genus *ANAX*

**1.** Sides of thorax purplish brown, reddish
brown or dark brown (exceptionally
green) **A**.
♂ Cercoids with a small external apical
tooth **B**.
♀ Two narrow tooth-like occipital
tubercles present **C**.
. . . . . . . . . . . . . . . . . . . ***Anax parthenope***

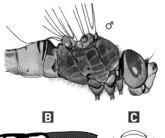
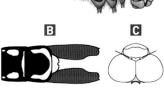

**NORTH:** JFMAM**JJAS**OND
**SOUTH:** JFMAM**JJAS**OND

**DISTRIBUTION:** Mediterranean area,
more scarce further north, although
its range is expanding.

**HABITAT:** sunny, stagnant waters,
sometimes brackish waters; prefers
waters with extensive submerged
vegetation, surrounded by reedbeds.

**CONFUSION SPECIES:** *Hemianax
ephippiger* and *A. imperator*.

**1'.** Sides of thorax green **A A**. Cercoids of
males and occiput of females variable.
. . . . . . . . . . . . . . . . . . . . . . . . . . . . . . . **2**

**2.** ♂ Cercoids without a tooth or external
apical spine **B**. Supra-anal plate reaching
approximately one-third the length of
cercoids and linear at its apex.
♀ No occipital tubercles **C**.
. . . . . . . . . . . . . . . . . . . . ***Anax imperator***

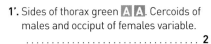
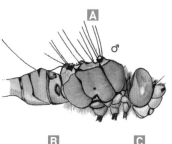

**NORTH:** JFMAM**JJASON**D
**SOUTH:** JFMAM**JJASON**D

**DISTRIBUTION:** less common further
north but its range is expanding.

**HABITAT:** sunny, stagnant and weakly
flowing waters, sometimes even
brackish, acidic or polluted waters;
up to 1,600m.

**CONFUSION SPECIES:** can be confused
with *A. parthenope*, which is smaller,
but also with *A. junius*.

**2'.** ♂ Cercoids with a strong external apical
spine **B**. Supra-anal plate reaching only
one-sixth the length of the cercoids and
more or less emarginate.
♀ Two pyramidal occipital tubercles
present **C**.
. . . ***Anax junius*** (exceptional American
migrant)

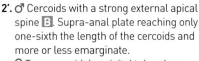
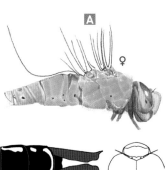

**DISTRIBUTION:** a Central and North
American species that very rarely
reaches Europe's Atlantic coast durir
autumnal storms. No reproduction
has been observed in our region.

**HABITAT:** fresh and brackish stagnant
waters; slow-flowing rivers.

**CONFUSION SPECIES:** *A. imperator*.

## *Anax parthenope* | Lesser Emperor

● Eyes touching along a line

● ♂ ♀ Blue markings clearly visible on S2 and S3 in males and young females

♂ abdomen: 46–53mm

♀ abdomen: 46–53mm

## *Anax imperator* | Blue Emperor/Emperor Dragonfly

● Eyes touching along a line

♂ abdomen: 53–64mm

♀ abdomen: 49–61mm

## *Anax junius* | Green Darner

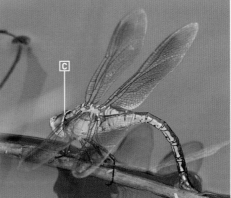

● Eyes touching along a line

♂ abdomen: 44–50mm

♀ abdomen: 46–56mm

Claudine and Pierre Guezennec

# Family Gomphidae

**1.** Discoidal cells divided into three or four small cells by secondary veins **A**. Abdominal segments 7 and 8 of males and females with large, prominent foliations **B**.
................ *Lindenia tetraphylla*

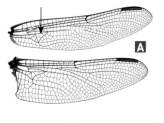

**1'.** Discoidal cells uninterrupted, without secondary veins ................... **2**

**2.** Hindwings without anal field; areas starting at rear edge of wing reaching anal vein **A**.
................................... **3**

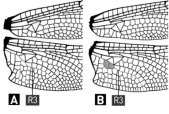

**2'.** Hindwings with anal field at base; areas starting at rear edge of wing not reaching anal vein **B**........................ **4**

nodus

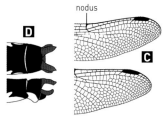

**3.** No prominent foliations at tip of abdomen. Seven to 12 transverse veins between nodus and pterostigma **C**.
♂ Cercoids equal in length to 10th abdominal segment **D**.
....... **Gomphus** and **Stylurus** (p. 70)

nodus

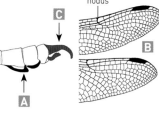

**3'.** Prominent foliations present on sides of abdominal segments 8 and 9, large and clearly visible in ♂ **A**, small and not very visible in ♀. Five or 6 transverse veins between nodus and pterostigma **B**.
♂ Cercoids at least twice as long as 10th abdominal segment **C**; supra-anal plate short, with a distinctive curvature in the shape of a 'double S'.
............... *Paragomphus genei*

**4.** ♂ Cercoids longer than 10th abdominal segment, curved downwards at a right angle **E**; supra-anal plate strongly curved upwards.
♀ Vulvar plate divided into two rounded lobes **F**.
Occiput without denticulated tubercles (no tubercles or smooth tubercles).
.............. **Onychogomphus** (p. 74)

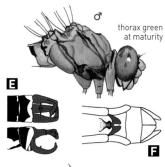

♂ thorax green at maturity

**4'.** ♂ Cercoids equal in length to 10th abdominal segment, not curved downwards at a right angle; supra-anal plate strongly curved upwards **A**.
♀ Vulvar plate bifid and terminated in two sharp points **B**. Occiput with two denticulated tubercles.
............. *Ophiogomphus cecilia*

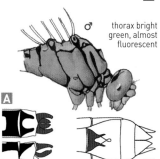

♂ thorax bright green, almost fluorescent

**NORTH AND SOUTH:** JFMAM**JJAS**OND

**DISTRIBUTION:** rare in the south-west of the Iberian Peninsula, Corsica, Sardinia, Sicily and southern Italy.

**HABITAT:** flowing waters, including hillside streams and small unpolluted semi-arid and sandy coastal rivers; below 500m.

**CONFUSION SPECIES:** *Lindenia tetraphyll*

**NORTH:** JFMAM**JJAS**OND
**SOUTH:** JFMAM**JJAS**OND

**DISTRIBUTION:** eastern Europe, with isolated populations in France and Italy.

**HABITAT:** flowing waters with sandy beds in lowland areas, from large rivers to small streams. Often found in the company of *Stylurus flavipes*.

**CONFUSION SPECIES:** for ♀, with *Aeshn cyanea*.

## Lindenia tetraphylla | Bladetail

- Eyes distinctly separate
- ♀ Base of the abdomen reddish

♂ abdomen : 48–57mm

♀ abdomen : 47–55mm

## Paragomphus genei | Green Hooktail

- Eyes distinctly separate
- Thorax green

♂ abdomen : 30–36mm

♀ abdomen : 28–35mm

## Ophiogomphus cecilia | Green Snaketail/Green Clubtail

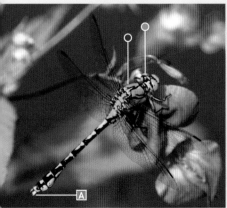

- Eyes green and distinctly separate
- Thorax green

♂ abdomen : 37–40mm

♀ abdomen : 37–42mm

## Genera *GOMPHUS* and *STYLURUS*

**1.** Tibiae entirely black, or at most with a short vestigial yellow line at base. Femurs black or black and yellow.
. . . . . . . . . . . . . . . . . . . . . . . . . . . . . . . **2**

**1'.** Legs streaked with yellow and black over entire length.
. . . . . . . . . . . . . . . . . . . . . . . . . . . . . . **3**

**2.** Abdomen strongly widened from seventh to ninth segments, marked with a clear longitudinal mediodorsal line on the first seven segments. Segments 8–10 entirely black dorsally **A**.
♂ Cercoids end in an acute short point, without external lateral teeth **B**.
♀ Vulvar plate bifid, very wide at base and with concave sides **C**.
. . . . . . . . . . . *Gomphus vulgatissimus*

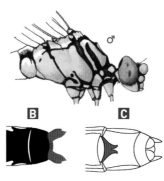

**North:** JFMA**MJJAS**OND
**South:** JFMA**MJJAS**OND

**Distribution:** common in many regions, especially in central Europe, but in sharp decline due to water pollution.

**Habitat:** sunny, flowing waters surrounded by trees and shrubs up to 1,000m. Sometimes on large turbuled lakes and well-oxygenated gravel pits

**Confusion species:** none.

**2'.** Abdomen only slightly widened towards its extremity, marked with a clear longitudinal mediodorsal line over its entire length **A**.
♂ Cercoids with strong external lateral teeth **B**.
♀ Vulva plate notched over at least half its length, with convex or straight sides **C**.
. . . . . . . . . . . . . . . . . *Gomphus graslinii*

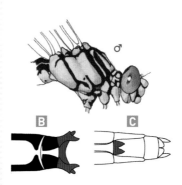

**North and South:** JFMA**MJJAS**OND

**Distribution:** endemic to south-western France and the Iberian Peninsula.

**Habitat:** large, calm rivers and small streams; also reservoirs; up to 300m

**Confusion species:** *G. simillimus* (females especially), which can be found in the same localities.

**3.** Abdomen slightly widened at its extremity, marked with a yellow mediodorsal line over its entire length **A**.
Median and antehumeral black bands of thorax curved in opposite directions and joining to form an oval pattern **D**.
♂ Cercoids quite long and curved inwards, without external lateral teeth **B**.
♀ Vulvar plate notched over its entire length **C**.
. . . . . . . . . . . . . . . . . . *Stylurus flavipes*

**3'.** Black bands of thorax almost straight; antehumeral and humeral bands parallel **D**.
♂ Cercoids rather short and straight in dorsal view. . . . . . . . . . . . . . . . . . . . . . . **4**

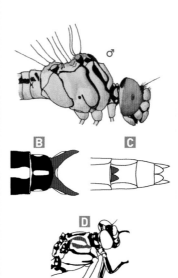

**North:** JFMA**MJJAS**OND
**South:** JFMA**MJJAS**OND

**Distribution:** common in eastern Eur and on large rivers such as the Loire, Rhône and Danube.

**Habitat:** large, natural rivers with bed sand, silt or mud; low altitudes.

**Confusion species:** *Gomphus graslinii G. simillimus*.

## *Gomphus vulgatissimus* | Common Clubtail/Club-tailed Dragonfly

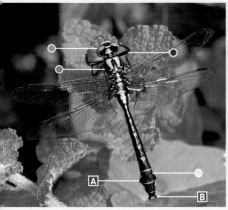
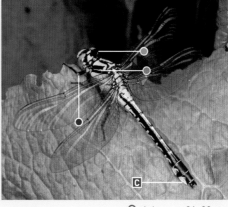

- Eyes distinctly separate
- Legs entirely black (key trait)
- Thin yellow antehumeral bands on the thorax
- Last segments black dorsally

♂ abdomen : 32–38mm

♀ abdomen : 31–39mm

## *Gomphus graslinii* | Pronged Clubtail

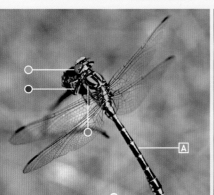

- Eyes distinctly separate
- Legs black with a yellow band on the femurs
- Thin yellow antehumeral bands on the thorax
- Yellow line on the last segments

♂ abdomen : 31–37mm

♀ abdomen : 31–38mm

## *Stylurus flavipes* | River Clubtail

- Eyes distinctly separate
- Legs black and yellow along all their length
- Large yellow antehumeral bands on the thorax

♂ abdomen : 32–40mm

♀ abdomen : 35–42mm

**4.** Median suture of thorax sides highlighted in black over its entire length 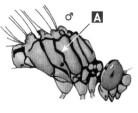. Ante-humeral and humeral black bands of thorax no wider than the yellow band they enclose **B**.
♀ Vulvar plate bilobed, shorter than half of ninth abdominal segment **C**.
. . . . . . . . . . . . . . .*Gomphus pulchellus*

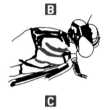

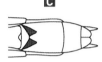

**NORTH:** JFMA**MJJAS**OND
**SOUTH:** JFMA**MJJAS**OND

**DISTRIBUTION:** generally common in south-west Europe, with a relatively recent expansion north and west of the Rhine.

**HABITAT:** various flowing and stagnant waters of any size; acid or alkaline waters, rich in aquatic vegetation or not, sometimes even in peat bogs.

**CONFUSION SPECIES:** *G. simillimus.*

**4'.** Median suture of thorax sides highlighted in black only at base 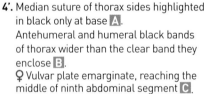.
Antehumeral and humeral black bands of thorax wider than the clear band they enclose **B**.
♀ Vulvar plate emarginate, reaching the middle of ninth abdominal segment **C**.
. . . . . . . . . . . . . . .*Gomphus simillimus*

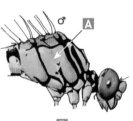

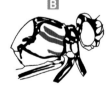

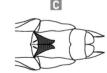

**NORTH:** JFMAMJJASOND
**SOUTH:** JFMAMJJASOND

**DISTRIBUTION:** generally common in France, Spain and Portugal, and occurring as a vagrant in neighbouring countries; declining.

**HABITAT:** flowing waters, including streams and large rivers, and sometimes in areas fed by groundwater; up to 500m.

**CONFUSION SPECIES:** *G. pulchellus.*

● Eyes distinctly separate

● Legs with black and yellow stripes

● Abdomen not widened at the tip

♂ abdomen: 34–38mm

♀ abdomen: 34–38mm

Side of the thorax

● Eyes distinctly separate

● Legs with black and yellow stripes

● Abdomen slightly widened at the tip

♂ abdomen: 33–36mm

♀ abdomen: 34–38mm

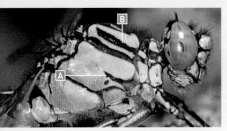

Side of the thorax

## Genus *ONYCHOGOMPHUS*

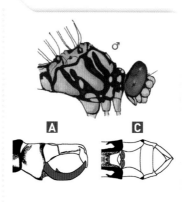

**1.** ♂ Branches of the supra-anal plate not differentiated at apex **A**. Cercoids not bifid at tip, in the shape of strongly hooked claws **B**.
♀ Vulvar plate formed by two converging blades, these not reaching a quarter of length of ninth abdominal segment **C**.
No tubercles on occiput behind eyes.
. . . . . . . . . . *Onychogomphus uncatus*

**NORTH AND SOUTH:** JFMAMJJASOND

**DISTRIBUTION:** generally common in th Iberian Peninsula, south and southwest France, and western Italy.

**HABITAT:** well-oxygenated white waters, especially in fast-flowing basin heads with pristine water; below 800m.

**CONFUSION SPECIES:** *O. forcipatus*.

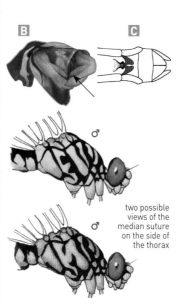

**1'.** ♂ Branches of supra-anal plate differentiated at apex into a subterminal ventral protuberance and a terminal dorsal projection **A A**. Cercoids bifid at tip, with inner lobe smaller than outer lobe **B B**.
♀ Vulvar plate bilobed, extending beyond middle of ninth abdominal segment **C**. Two smooth tubercles present on occiput.
. . . . . . . . *Onychogomphus forcipatus*
(two subspecies)

two possible views of the median suture on the side of the thorax

**NORTH:** JFMAMJJASOND
**SOUTH:** JFMAMJJASOND

**DISTRIBUTION:** fairly common across Continental Europe. Absent from Britain and Ireland.

**HABITAT:** flowing waters, including lively streams, rivers and canals; sometimes lake waters, including large, turbulent and well-oxygenated lakes and gravel pits; up to 1,200m.

**CONFUSION SPECIES:** *O. uncatus*.

• Apex of supra-anal plate continuing its general curve or forming only a strongly obtuse angle of 120–170°, with its curve directed upwards or backwards.
. . . . . . . . . . *Onychogomphus forcipatus*
*forcipatus*

**NORTH:** JFMAMJJASOND
**SOUTH:** JFMAMJJASOND

**DISTRIBUTION:** Iberian Peninsula, southern France and north-west Italy.

**HABITAT:** flowing waters, and sometimes in lakes; up to 1,200m.

**CONFUSION SPECIES:** *O. uncatus*.

• Apex of supra-anal plate bent at a right angle, forming an angle of 30–120° with it, and directed forward.
. . . . . . . . . . *Onychogomphus forcipatus*
*unguiculatus*

## *Onychogomphus uncatus* | Large Pincertail/Blue-eyed Hooktail

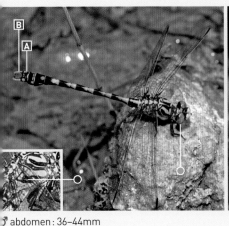

● Eyes distinctly separate

● Large, linked lateral black bands (in contrast with *O. forcipatus*)

♂ abdomen: 36–44mm

♀ abdomen: 34–39mm

## *Onychogomphus forcipatus forcipatus* | Small Pincertail/Green-eyed Hooktail

● Eyes distinctly separate

● Thin lateral black band (in contrast with *O. uncatus*)

♂ abdomen: 31–40mm

♀ abdomen: 31–39mm

## *Onychogomphus forcipatus unguiculatus* | (Southern) Small Pincertail/Green-eyed Hooktail

● Eyes distinctly separate

● Thin lateral black band (in contrast with *O. uncatus*)

♂ abdomen: 31–40mm

♀ abdomen: 31–39mm

1. Discoidal cells longitudinal on both fore-wings and hindwings A. General colour black and yellow.
   ................................... 2

1'. Discoidal cells transverse on forewings and longitudinal on hindwings (costal side shorter than the two other sides on fore-wings, longer on hindwings) A. General colour black and yellow on abdomen. Thorax metallic green marked with yellow humeral and lateral bands B.
   ............. ***Macromia splendens***

NORTH AND SOUTH: JFMAMJJASOND

DISTRIBUTION: rare and local in south France and the Iberian Peninsula.

HABITAT: calm areas of large rivers, hydroelectric reservoirs and small streams with relatively deep basins; below 500m altitude.

CONFUSION SPECIES: *Cordulegaster* sp

2. Sides of first abdominal segment with a yellow club-shaped sub-rectilinear patch in upper part A. Occipital triangle black and only slightly swollen.
   ♂ Cercoids spreading from base, with two ventral teeth visible in profile view; supra-anal plate reaching two-thirds of cercoids B.
   ♀ Ovipositor entirely black.
   ............ ***Cordulegaster bidentata***

NORTH: JFMAMJJASOND
SOUTH: JFMAMJJASOND

DISTRIBUTION: endemic to Continenta Europe, regionally common from cer France eastwards to Bulgaria.

HABITAT: mainly in open deciduous submontane and montane woodland up to 1,400m in headwaters.

CONFUSION SPECIES: *C. boltonii* and *Macromia splendens*.

2'. Sides of first abdominal segment with a C-shaped yellow lateral patch along posterior and lower edges A. Occipital triangle yellow and distinctly swollen.
   ♂ Cercoids held close together at base, with a single ventral tooth visible in profile view; supra-anal plate not exceeding half of cercoids B.
   ♀ Ovipositor black with two yellow lateral patches at base.
   ............. ***Cordulegaster boltonii***
   (two geographical forms formerly considered subspecies, but not currently presenting any genetic differentiation)

• Yellow mediodorsal spots on abdomen small and separate, equal to one-fifth the length of the segments C. Frons yellow with a strongly marked transverse black line along superior carina, this sometimes absent in atypical individuals.
   ..... *Cordulegaster boltonii boltonii*

• Yellow mediodorsal spots on abdomen large, merging into continuous rings, often reaching one-quarter to one-third the length of segments 3–8 C'. Frons entirely yellow or with a small, reduced black line.
   ............... *Cordulegaster boltonii immaculifrons*

*Cordulegaster boltonii boltonii*     *Cordulegaster b immaculifro*

NORTH: JFMAMJJASOND
SOUTH: JFMAMJJASOND

DISTRIBUTION: widespread across the region as far north as the Arctic Circ and locally common.

HABITAT: sandy river beds in submontane and montane areas, whe it may form large localised population

CONFUSION SPECIES: *C. bidentata* and *Macromia splendens*, both of which are similar in size and have a similar flight style.

## Macromia splendens | Splendid Cruiser

- Eyes touching at a single point. Two distinctive symmetrical yellow spots on the frons

- ♂ ♀ Large yellow patch on segment 7

♂ abdomen: 49–55mm

♀ abdomen: 48–54mm

## Cordulegaster bidentata | Sombre Goldenring/Two-toothed Goldenring

- ♂ ♀ Eyes touching at a single point; occipital triangle black, not inflated

- Anal triangle usually made of 3 cells

♂ abdomen: 50–59mm

♀ abdomen: 56–65mm

## Cordulegaster boltonii boltonii | Common Goldenring/Golden-ringed Dragonfly

- ♂ ♀ Eyes touching at a single point; occipital triangle yellow, inflated and very distinct

- Anal triangle usually made of 4–6 cells

♂ abdomen: 45–61mm

♀ abdomen: 65–69mm

**1.** General colour of thorax and abdomen
yellow and black. Hindwings with large
black patch at base **A**.
♂ Abdomen tapered and with V-shaped
cercoids in **B**.
♀ Vulvar plate bifid and reaching middle
of 10th abdominal segment **C**.
................*Epitheca bimaculata*

**1'.** Thorax metallic green, with or without
yellow lines. Abdomen metallic green to
blackish, with or without yellow spots.
Wings without black patch at base **A**.
♀ Vulvar plate not as above .......... **2**

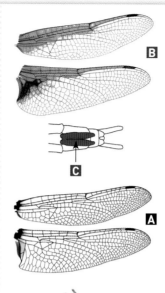

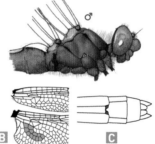

**NORTH:** JFMAM**JJAS**OND
**SOUTH:** JFMAM**JJAS**OND

**DISTRIBUTION:** central and north-
eastern Continental Europe.

**HABITAT:** stagnant waters in semi-
forested areas, including ponds and
old gravel pits, with or without floating
hydrophytes.

**CONFUSION SPECIES:** *Libellula
quadrimaculata*.

**2.** Anal field of hindwings a simple elongated
curve **B**. Abdomen metallic green to black
with distinct, elongated yellow spots present only on the mediodorsal side **A**.
♂ 10th segment with a clear dorsal crest.
♀ Vulvar plate very short and reduced to
two lobes **C**.
................. *Oxygastra curtisii*

**2'.** Anal field of hindwings in the shape of a
foot **B**. Abdomen metallic green to black,
sometimes marked with lateral yellow
spots **A**.
♂ 10th segment without dorsal crest.
♀ Well-developed vulvar plate.
................................... **3**

**NORTH:** JFMAM**JJAS**OND
**SOUTH:** JFMAM**JJAS**OND

**DISTRIBUTION:** south-west Europe,
including western Italy, much of
France and the Iberian Peninsula.
Extinct in Britain and the Netherlands.

**HABITAT:** mainly flowing waters,
especially the calm parts of large
rivers with wooded banks; sometimes
in stagnant waters; below 800m.

**CONFUSION SPECIES:** unlikely.

**3.** Hindwings with a single cubito-anal vein,
located between discoidal cell and base
of wing **B**.
♂ Supra-anal plate doubly bifid **D**.
♀ Vulvar plate deeply notched **C**.
.................... *Cordulia aenea*

**3'.** Hindwings with two transverse cubito-
anal veins, located between discoidal
cell and base of wing **C**.
♂ Supra-anal plate entire **D**.
♀ Vulvar plate entire, emarginate or
notched.
................ *Somatochlora* (p. 80)

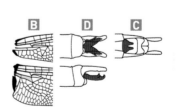

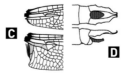

**NORTH:** JFMAM**JJAS**OND
**SOUTH:** JFMAM**JJAS**OND

**DISTRIBUTION:** very common across
parts of the region; absent from the
Iberian Peninsula.

**HABITAT:** all kinds of stagnant waters
and acid or alkaline mountain bogs;
exceptionally on canals and in the
calm parts of rivers.

**CONFUSION SPECIES:** resembles
*Somatochlora* spp.

## *Epitheca bimaculata* | Eurasian Baskettail/Two-spotted Dragonfly

- ● Eyes touching at one spot

- ● Nodus of all four wings lacking a black patch (by contrast with *L. quadrimaculata*)

♂ abdomen : 37–42mm

♀ abdomen : 37–43mm

## *Oxygastra curtisii* | Orange-spotted Emerald

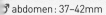

- ● Eyes green, touching at one spot

- ● ♂ ♀ Thorax entirely metallic green

♂ abdomen : 33–39mm

♀ abdomen : 34–35mm

## *Cordulia aenea* | Downy Emerald

- ● Eyes green, touching at one spot

- ● ♂ ♀ Thorax entirely metallic green

- ● Abdomen widened and club-shaped in males (by contrast with *Somatochlora metallica*)

♂ abdomen : 32–39mm

♀ abdomen : 30–39mm

## Genus *SOMATOCHLORA*

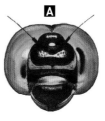

**1.** Front of frons marked with a transverse yellow line connecting its two sides .
♀ Vulvar plate very long, triangular and forming a marked right angle with abdomen.
.................................. **2**

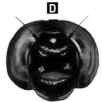

**1'.** Front of frons lacking transverse yellow line .
♀ Vulvar plate not as above .......... **3**

**NORTH:** JFMAMJJASOND
**SOUTH:** JFMAMJJASOND

**DISTRIBUTION:** common across much of northern Europe.

**HABITAT:** breeds in ponds, lakes and peat bogs; sometimes in slow-flowing rivers and fast-moving streams; restricted to southern mountain lake

**CONFUSION SPECIES:** *Cordulia aenea* an the four other *Somatochlora* species.

**2.** Sides of thorax entirely metallic green, without a yellow patch **A**.
Pterostigmas brown **B**. Hindwings with a yellow spot at base along membranule **C**.
Abdomen of ♀ with or without latero-dorsal yellow spots on second and third segments.
............. ***Somatochlora metallica***

**NORTH:** JFMAMJJASOND
**SOUTH:** JFMAMJJASOND

**DISTRIBUTION:** south-east Europe, reaching north to Austria and west to France.

**HABITAT:** breeds in running water, often forest streams at low altitudes and open areas at higher altitudes.

**CONFUSION SPECIES:** *Cordulia aenea* an the four other *Somatochlora* species.

**2'.** Sides of thorax marked with one or two yellow patches in both sexes **A**.
Pterostigmas black **B**. Base of wings translucent along membranule **C**.
Abdomen of ♀ with distinct yellow dorsolateral spots on second and third segments **D**.
........ ***Somatochlora meridionalis***

**3.** Sides of thorax marked with two yellow bands, which darken and almost fade entirely with age **A**. Abdomen with yellow lateral patches on segments 1–8 (9) **D**.
♂ Cercoids almost linear and smooth in dorsal view; lateral teeth visible only in profile view **B**.
♀ Vulvar plate emarginate **C**.
....... ***Somatochlora flavomaculata***

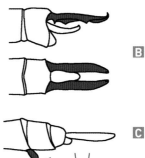

**3'.** Sides of thorax entirely metallic green. Abdomen with, at most, yellow spots at base.
♂ Cercoids not as above.
♀ Vulvar plate sub-rectilinear or convex at tip.
.................................. **4**

**NORTH:** JFMAMJJASOND
**SOUTH:** JFMAMJJASOND

**DISTRIBUTION:** north-east Europe; generally uncommon and declining, despite large local populations.

**HABITAT:** ponds, lakes, old gravel pits dead arms of large rivers; mesotrop marshes and lowland peatlands; up to 1,300m.

**CONFUSION SPECIES:** *Cordulia aenea* an the four other *Somatochlora* species.

## *Somatochlora metallica* | Brilliant Emerald

- Eyes green, touching at one spot; transverse, lateral yellow band on the frons of the forehead

- Abdomen green, elongated and fusiform (by contrast with *C. aenea*)

Thomas Roussel (Biotope)

♂ abdomen: 37–40mm

♀ abdomen: 40–44mm

## *Somatochlora meridionalis* | Balkan Emerald

- Eyes green, touching at one spot; transverse, lateral yellow band on the frons of the forehead

- Abdomen green, elongated and fusiform, marked with one or two yellow spots on S2 and S3 (by contrast with *C. aenea*)

♂ abdomen: 37–40mm

♀ abdomen: 40–44mm

## *Somatochlora flavomaculata* | Yellow-spotted Emerald

- Eyes green, touching at one spot

Alain Cochet

♂ abdomen: 34–38mm

♀ abdomen: 34–43mm

4. Two transverse cubito-anal veins on fore-wings, located between discoidal cell and base of wing **D**. Abdomen lacking large rounded laterodorsal yellow spots on the third segment **A**.
♂ Cercoids sublinear, with an angular external edge in dorsal view **B**.
♀ Vulvar plate short and erect in relation to abdomen **C**.
. . . . . . . . . . . . **Somatochlora alpestris**

4'. A single transverse cubito-anal vein on forewings, located between discoidal cell and base of wing **D**.
Abdomen with two large rounded yellow laterodorsal spots on third segment in ♀ **A**.
♂ Cercoids pincer-shaped and smooth in dorsal view **B**.
♀ Vulvar plate appressed to ninth abdominal segment **C**.
. . . . . . . . . . . . . **Somatochlora arctica**

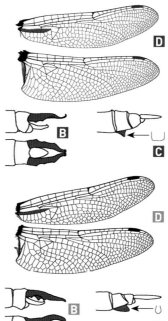

**NORTH AND SOUTH:** JFMAMJJASOND

**DISTRIBUTION:** boreo-alpine parts of t region, above 800m in central Europ

**HABITAT:** *Sphagnum* peat bogs and peat ponds; up to 2,250m.

**CONFUSION SPECIES:** *Cordulia aenea* ar the four *Somatochlora* species.

# Family Libellulidae

1. Base of hindwings with opaque brown or blackish patch, sometimes surrounded by a translucent saffron-yellow patch . . . . **2**

1'. Base of hindwings translucent or saffron-coloured . . . . . . . . . . . . . . . . . . **3**

2. More than 12 transverse antenodal veins on forewings. Pterostigmas more than 3.5 times longer than wide **A**.
. . . . . . . . . . . . . . . . . . . . . **Libellula** (p. 88)

2'. Fewer than 10 transverse antenodal veins on forewings. Pterostigmas less than 3.5 times longer than wide **B**.
. . . . . . . . . . . . . . **Leucorrhinia** (p. 100)

3. Five or six transverse antenodal veins on forewings; last antenodal vein complete **C**.
. . . . . . . . . **Selysiothemis nigra** (p. 86)

3'. Seven to 13 transverse antenodal veins on forewings, the last complete or incomplete . . . . . . . . . . . . . . . . . . . . . . . **4**

4. Post-discoidal area of forewings widening from middle **D**.
. . . . . . . . . . . . . . . . . . . . . . . . . . . . . . . . . **5**

4'. Post-discoidal area of forewings of constant width, or narrowing from middle **E**.
. . . . . . . . . . . . . . . . . . . . . . . . . . . . . . . **6**

**NORTH:** JFMAMJJASOND
**SOUTH:** JFMAMJJASOND

**DISTRIBUTION:** north-east Europe, especially Fennoscandia; generally located in mountain ranges and their foothills further south.

**HABITAT:** acid and neutral peat bogs; sea level to >2,000m.

**CONFUSION SPECIES:** *Cordulia aenea* ar the four *Somatochlora* species.

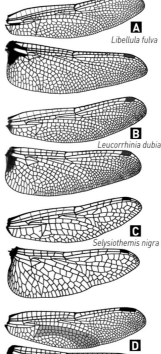

*Libellula fulva*

*Leucorrhinia dubia*

*Selysiothemis nigra*

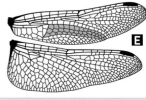

- Eyes green, touching at one spot
- ♀ Abdomen robust
- Thorax without yellow lateral bands
- White junction between S2 and S3 clearly visible

♂ abdomen: 32–34mm

♀ abdomen: 31–35mm

- Eyes green, touching at one spot
- Base of the abdomen strongly constricted

♂ abdomen: 30–36mm

♀ abdomen: 30–37mm

**5.** Last transverse antenodal vein incomplete on forewings 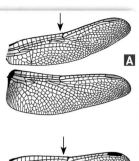. Abdomen large and depressed in both sexes.
♂ Abdomen scarlet-red at maturity **B**.
♀ Abdomen yellowish brown or bright red, not criss-crossed with black **C**.
...................... ***Crocothemis erythraea***

**NORTH:** JFMAMJJAS**OND**
**SOUTH:** JFM**AMJJAS**OND

**DISTRIBUTION:** common to very common across south and central Europe, and expanding northwards.

**HABITAT:** stagnant and weakly flowing waters, even brackish or notably eutrophic waters; up to >1,300m.

**CONFUSION SPECIES:** none.

**5'.** Last transverse antenodal vein complete on all four wings 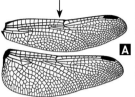.
♂ Abdomen blue at maturity.
♀ Abdomen yellow-brown, distinctly striped or criss-crossed with black, or sometimes with bluish markings.
...................... ***Orthetrum*** (p. 90)

**6.** Forewings with 6.5–8.5 transverse antenodal veins **B**. ...................................7

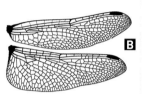

**6'.** At least 9.5 transverse antenodal veins on fore-wings **C**. ................................ **8**

*Brachythemis impartita*

**NORTH AND SOUTH:** JFM**AMJJASO**ND

**DISTRIBUTION:** in our region occurs in the Iberian Peninsula, Sardinia, Corsica and southern Italy; expanding northwards.

**HABITAT:** stagnant and weakly flowing waters, with a variable water level and bare sandy banks.

**CONFUSION SPECIES:** none.

**7.** Pterostigmas bicoloured, white with a dark brown tip. Wings with a broad brown stripe, this not encompassing pterostigma but sometimes touching it. ......... ***Brachythemis impartita***

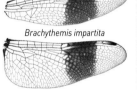

**7'.** Pterostigmas of a single colour. Wings without a brown band, or with a brown band encompassing part of pterostigma ........***Sympetrum*** (p. 94)

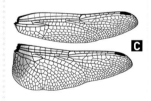

**8.** Pterostigmas roughly same size on all four wings **D**. Discoidal cells at almost same level on all four wings **E**.
......................... ***Trithemis*** (p. 86)

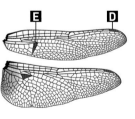

**DISTRIBUTION:** a tropical and North American species, frequent in our region only in Sicily, Cyprus and southern Turkey; scattered records elsewhere across the eastern Mediterranean and central and eastern Europe.

**HABITAT:** stagnant waters, especially temporary waters and sometimes brackish waters; more rarely in weakly flowing waters.

**CONFUSION SPECIES:** none.

**8'.** Pterostigmas markedly longer on forewings than on hindwings **A**. Discoidal cells on forewings located beyond those on hindwings **B**. Abdomen yellowish to red **C**.
...................... ***Pantala flavescens***

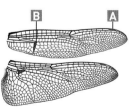

## *Crocothemis erythraea* | Broad Scarlet/Scarlet Darter

- A clearly distinct yellow patch at the base of the hindwings
- Thorax reddish-brown, legs red

♂ abdomen: 19–31mm

♀ abdomen: 18–29mm

## *Brachythemis impartita* | Northern Banded Groundling

- Large brown band crossing the pterostigma
- Pterostigma bicoloured
- ♂ Body dark brown at maturity
- ♀ Body beige and black
- ♀ Abdominal appendages light beige, the tip of superior appendages blackish

♂ abdomen: 16–23mm

♀ abdomen: 16–24mm

## *Pantala flavescens* | Globe Skimmer/Globe Wanderer/Wandering Glider

- Hindwings particularly large at base

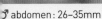
♂ abdomen: 26–35mm

♀ abdomen: 26–37mm

## Genus **TRITHEMIS**

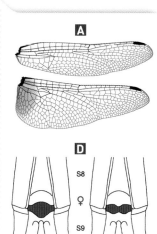

**1.** Base of hindwings marked with a yellow spot, this reaching at most proximal side of discoidal cell **A**. Frons notched with a shallow groove, entirely tan to metallic violet above in ♂ **B**.
♂ Abdomen more or less purplish red depending on age.
♀ Abdomen variable, with yellow, brown and black lines on carinae **C**. Vulvar plate rounded-convex or sinuous at tip **D**.
. . . . . . . . . . . . . . . . . . . . . ***Trithemis annulata***

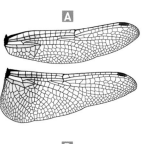

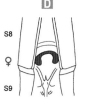

**1'.** Base of hindwings marked with a large yellow spot encompassing the discoidal cell in ♂, and sometimes also in ♀ **A**. Frons notched with a very deep groove; mainly bright red in ♂, yellow-brown in ♀, without significant metallic aspect **B**.
♂ Abdomen bright red at maturity.
♀ Abdomen yellow to orange-brown with variable dorsolateral black lines between lateral and dorsal carinae **C**. Vulvar plate bearing a deep U-shaped invagination with parallel edges **D** . . . . . . . . ***Trithemis kirbyi***

**Note on *Trithemis kirbyi*:** this Afrotropical species has been permanently established in the Iberian Peninsula since 2007, and reached the south of France in 2017 and 2018 during exceptional heat waves. There is no evidence that the species is currently breeding in France.

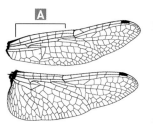

## Genus **SELYSIOTHEMIS**

**1.** Pale, loose venation, with five or six transverse antenodal veins on forewings; last antenodal vein is complete **A**.
. . . . . . . . . . . . . . . . . . . . ***Selysiothemis nigra***

**Note on *Selysiothemis nigra*:** widely distributed from central Asia to the Middle East, North Africa and southern Europe. Well-established populations throughout the Mediterranean Basin and south-west Asia are apparently becoming more dense due to changes in artificial habitats. However, the species is not expanding significantly outside of its original range. The establishment of the species in the north of Corsica appears to be confirmed, with three distinct localities identified from 2015 to 2017 and evidence of successful breeding.

**NORTH AND SOUTH:** JFMAMJJASOND

**DISTRIBUTION:** along the Mediterrane and expanding to the north; observe for the first time in mainland France and Italy in the 1990s, and reached Hungary in 2016.

**HABITAT:** stagnant and flowing water at low altitudes, including ponds, gravel pits, streams and rivers.

**CONFUSION SPECIES:** *Crocothemis erythraea*.

**HABITAT:** a versatile species, found in permanent or temporary streams and ponds, in fresh or brackish wate including concrete fountains in urba environments and artificial irrigation ponds in agricultural areas.

**CONFUSION SPECIES:** *T. annulata, Sympetrum flaveolum* and *S. fonscolombii*.

**NORTH AND SOUTH:** JFMAMJJASOND

**HABITAT:** permanent or temporary ponds, in fresh or brackish water, including abandoned gravel pits and rivers with variable flow.

**CONFUSION SPECIES:** *Diplacodes lefebvr* in the Iberian Peninsula and Turkey; *Sympetrum danae*.

## *Trithemis annulata* | Violet Dropwing/Violet-marked Darter

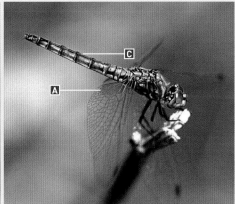

♂ abdomen : 17–29mm

♀ abdomen : 19–24mm

## *Trithemis kirbyi* | Orange-winged Dropwing

♂ abdomen : 19–23mm

♀ abdomen : 19–23mm

## *Selysiothemis nigra* | Black Pennant

♂ abdomen : 19–26mm

♀ abdomen : 17–25mm

## Genus *LIBELLULA*

nodus

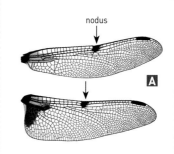

**1.** Nodus of all four wings with a blackish spot **A**. General colour yellowish brown to olive and black.
. . . . . . . . . . .***Libellula quadrimaculata***

**1'.** No blackish spot on nodus of wings.
. . . . . . . . . . . . . . . . . . . . . . . . . . . . . . . . . **2**

**North:** JFMAMJJASOND
**South:** JFMAMJJASOND

**Distribution:** frequent and abundant across much of the region.

**Habitat:** all kinds of stagnant waters, sometimes also weakly flowing waters; breeds up to an altitude of 2,200m.

**Confusion species:** *Epitheca bimaculata*.

**2.** Black spot at base of wings reaching discoidal cell at forewings, and encompassing discoidal cell on hindwings **A**. Top of eyes dark brown **B**.
♂ Abdomen stout and depressed, with yellow lateral lunules; abdomen mostly blue at maturity **C**.
♀ Abdomen stout and depressed, yellow or brown; without a broad dark mediodorsal band **D**.
. . . . . . . . . . . . . . . . .***Libellula depressa***

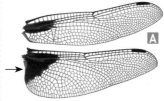

**North:** JFMAMJJASOND
**South:** JFMAMJJASOND

**Distribution:** a pioneer species that is extremely common in spring across m of the region, except at northernmost latitudes and high altitudes.

**Habitat:** stagnant and weakly flowing mesotrophic and eutrophic waters, e when brackish or slightly polluted.

**Confusion species:** none.

**2'.** Black spot at base of wings not reaching base of discoidal cell on forewings, and not encompassing discoidal cell on hindwings **A**. Top of eyes blue-grey **B**.
♂ Abdomen without yellow lateral lunules; abdomen blue with black tip at maturity **C**.
♀ Abdomen yellow with a broad black mediodorsal band on segments 4–10 **D**.
. . . . . . . . . . . . . . . . . . . . . .***Libellula fulva***

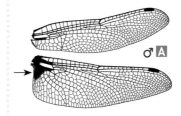

**North:** JFMAMJJASOND
**South:** JFMAMJJASOND

**Distribution:** frequent across the region in its preferred habitats.

**Habitat:** stagnant and weakly flowing mesotrophic or eutrophic waters.

**Confusion species:** *Orthetrum cancellatum*, but it has blue-grey eyes and a black spot at base of wings.

## *Libellula quadrimaculata* | Four-spotted Chaser

- ♂ ♀ Base of hindwings marked with a distinct black patch

- Abdomen tapered, with yellow lateral spots resembling those of *Epitheca bimaculata*

♂ abdomen : 27–32mm

♀ abdomen : 28–31mm

## *Libellula depressa* | Broad-bodied Chaser

♂ abdomen : 24–31mm

♀ abdomen : 21–29mm

## *Libellula fulva* | Scarce Chaser

- ♀ Black patches at the tip of the wings

- Black line at the base of the forewings

♂ abdomen : 26–29mm

♀ abdomen : 26–29mm

## Genus *ORTHETRUM*

**1.** Pterostigmas black.
.................................... **2**

**1'.** Pterostigmas yellow or brown.
.................................... **3**

**2.** Cercoids white in both sexes; 10th abdominal segment also white in ♀ **A**.
.............. *Orthetrum albistylum*

**2'.** Cercoids black in both sexes; 10th
abdominal segment never white **A**.
........... *Orthetrum cancellatum*

**3.** Base of abdomen swollen and bulbous **A**.
Width of abdomen constant from
fourth segment; abdomen longer than
hindwings **B**.
................ *Orthetrum trinacria*

**3.** Abdomen thickened only at base when
seen in lateral view, not bulbous; abdomen
shorter than hindwings.
.................................... **4**

**Note on *Orthetrum trinacria*:** an Afrotropical
species, now well established in south-west
Europe, especially in the southern half of the
Iberian Peninsula, Sardinia and Sicily. It appears
to have established itself in northern Corsica,
with around 30 individuals emerging at times,
and a number of exuviae spotted from 2012
to 2017 in the village of Lucciana. A nomadic
individual from Sardinia was also observed in
the Lavezzi Islands in 2013.

## Orthetrum albistylum | White-tailed Skimmer

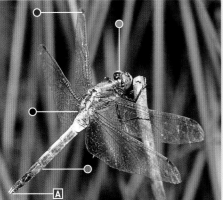

- Eyes touching at one point
- Pterostigmas black (as in *O. cancellatum*)
- Abdomen blue, or yellow and black
- ♀ 10th abdominal segment of females dorsally white (unlike in *O. cancellatum*)
- Base of the wings without a black patch

♂ abdomen: 32–34mm

♀ abdomen: 31–33mm

## Orthetrum cancellatum | Black-tailed Skimmer

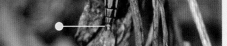

- Eyes touching at one point
- Pterostigmas black
- Abdomen blue, or yellow and black
- ♀ 10th abdominal segment of females black and yellow
- Base of the wings without a black patch (unlike in *Libellula fulva*)

♂ abdomen: 29–35mm

♀ abdomen: 27–34mm

## Orthetrum trinacria | Long Skimmer

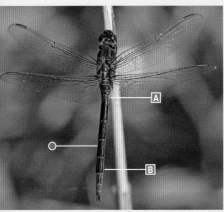

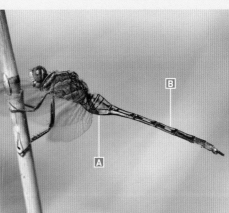

- ♂ General coloration: yellow and black, then blackish with highly variable blue pruinescence

♂ abdomen: 35–44mm

♀ abdomen: 36–42mm

**4.** Face white **A**. Two rows of cells between IR3 and Srv; the second row generally with more than five cells **B**.
. . . . . . . . . . . . . . **Orthetrum brunneum**

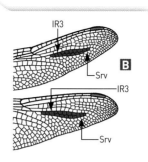

**NORTH:** JFMAMJJASOND
**SOUTH:** JFMAMJJASOND

**DISTRIBUTION:** common, especially around the Mediterranean. Absent fro Britain, Ireland and Fennoscandia.

**HABITAT:** stagnant and flowing waters; found only in lowlands in the north of i range, and up to 1,800m in the south.

**CONFUSION SPECIES:** *O. coerulescens*, but that species is generally slightly smaller and has a brown face.

**4'.** Face brown **A A**. One or two rows of cells between IR3 and Srv, the second row generally with no more than four cells **B B**.
. . . . . . . . . . **Orthetrum coerulescens**
(two subspecies)

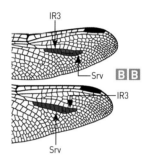

• Pterostigmas long, 3.3–4mm. Anterior lamina of male genitalia erect, projecting at right angles to base of abdomen and sub-cylindrical in profile view **C**. Andromorphic females (blue at maturity) rare.
. . . . . . . . . . **Orthetrum coerulescens coerulescens**

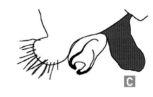

**NORTH:** JFMAMJJASOND
**SOUTH:** JFMAMJJASOND

**DISTRIBUTION:** common around the Mediterranean; more local in central and northern Europe.

**HABITAT:** stagnant and flowing waters including ponds, peat bogs, springs, small streams and flooded quarries; up to 1,600m.

**CONFUSION SPECIES:** *O. brunneum*, but that species has a white face.

• Pterostigmas short, <3.5mm. Anterior lamina of male genitalia erect, projecting obliquely to base of abdomen, and triangular in profile view **C**. Andromorphic females frequent.
. . . . **Orthetrum coerulescens anceps**

**Note:** It is impossible to distinguish the females of these two subspecies using the morphology of the vulvar plate.

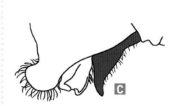

**NORTH AND SOUTH:** JFMAMJJASOND

**DISTRIBUTION:** Sardinia, Corsica and th southern Balkans.

**HABITAT:** stagnant and flowing waters

**CONFUSION SPECIES:** *O. brunneum*, but that species has a white face.

## *Orthetrum brunneum* | Southern Skimmer

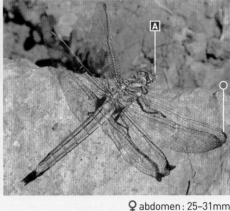

- Eyes touching at one spot; face white or pale blue
- ♂ Entirely blue
- Pterostigmas yellow-brown
- Base of the wings without a black patch
- Cercoids black

♂ abdomen : 25–34mm        ♀ abdomen : 25–31mm

## *Orthetrum coerulescens coerulescens* | Keeled Skimmer

- Eyes touching at one spot; face brown
- ♂ Thorax dark, rarely bluish
- Pterostigmas yellow to brown
- Base of the wings without a black patch
- Cercoids black
- Abdomen narrow, uniformly blue

♂ abdomen : 25–31mm        ♀ abdomen : 25–30mm

## *Orthetrum coerulescens anceps* | (Rambur's) Keeled Skimmer

- ♂ Thorax frequently powdery-blue

♂ abdomen : 23–30mm        ♀ abdomen : 23–29mm

## Genus *SYMPETRUM*

**1.** Wings crossed by a large brown band that meets pterostigmas .
........*Sympetrum pedemontanum*

**1'.** Wings without a dark band.
.................................. **2**

**2.** Femurs and tibiae entirely black.
.................................. **3**

**2'.** Femurs and/or tibiae yellow and black, or entirely yellow.
.................................. **5**

**3.** Thorax brown and black with two yellow bands on sides, which can darken and fade with age .
♂ Abdomen yellow and black, becoming entirely black with age 🅱.
♀ Characteristic black triangle on top of thorax 🅲.
................. *Sympetrum danae*

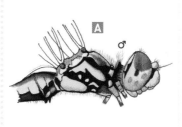

**3'.** Thorax yellow or brown with black thoracic sutures 🅰.
♂ Abdomen mainly red at maturity.
♀ No black triangle on top of thorax.
.................................. **4**

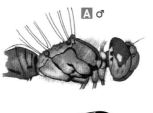

**4.** Four to five cells between Srv and rear edge of hindwings 🅱.
Abdomen not depressed dorsoventrally, but fusiform and constricted in segments 3–5 when seen in profile; abdomen marked with lateral black spots 🅲.
♂ Frons bright red at maturity 🅳.
.......... *Sympetrum sanguineum*

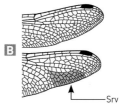
—Srv

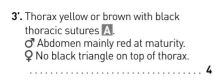

**NORTH:** JFMAMJJASOND
**SOUTH:** JFMAMJJASOND

**DISTRIBUTION:** highly localised in western Europe, becoming more common eastwards; nomadic individuals observed as far west as western France and the British Isles.

**HABITAT:** sunny, stagnant and weakly flowing mesotrophic and eutrophic waters; up to 700m.

**CONFUSION SPECIES:** none.

**NORTH:** JFMAMJJASOND
**SOUTH:** JFMAMJJASOND

**DISTRIBUTION:** boreal parts of the region and at higher altitudes further south.

**HABITAT:** stagnant acidic or temporar waters such as peat bogs; up to 2,000m.

**CONFUSION SPECIES:** *Selysiothemis nigr Diplacodes lefebvrii.*

**NORTH:** JFMAMJJASOND
**SOUTH:** JFMAMJJASOND

**DISTRIBUTION:** very common, especial at low altitudes; less common in Scotland and Fennoscandia, but expanding northwards.

**HABITAT:** stagnant, brackish or marked eutrophic waters surrounded by reedbeds and sedge beds; mainly in lowlands, locally up to 1,700m.

**CONFUSION SPECIES:** *S. depressiusculur*

● Eyes touching at one point

● Pterostigmas uniformly red or yellow

♂ abdomen : 18–23mm

♀ abdomen : 18–24mm

● Eyes bicoloured, touching at one point

● Legs entirely black

♂ abdomen : 20–26mm

♀ abdomen : 18–26mm

● Eyes touching at one point

● Legs entirely black

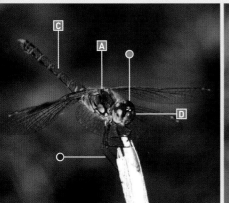

♂ abdomen : 20–26mm

♀ abdomen : 21–26mm

**4'.** Five to seven cells between Srv and rear edge of hindwings 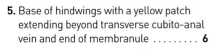.
Abdomen not constricted in centre when seen in profile view; abdomen almost depressed in ♂, sub-cylindrical in ♀, marked with characteristic cuneiform black patches on sides **B**.
Frons remaining brownish at maturity in both sexes **C**.
..... *Sympetrum depressiusculum*

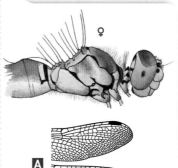

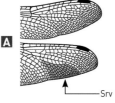

**5.** Base of hindwings with a yellow patch extending beyond transverse cubito-anal vein and end of membranule ......... **6**

Srv

**5'.** Base of hindwings translucent or with a defined yellow spot not exceeding transverse cubito-anal vein ............... **7**

**NORTH AND SOUTH:** JFMAMJJASOND

**DISTRIBUTION:** central and eastern Europe; globally decreasing, even if large populations remain.

**HABITAT:** shallow, sunny, stagnant waters, often temporary waters or th with variable water level, covered wi vegetation; sometimes also on swar riverbanks; low altitudes.

**CONFUSION SPECIES:** *S. sanguineum*.

**6.** Basal yellow spot on hindwings reaching at most base of discoidal cell **A**.
Postero-inferior surface of eyes bluish grey at maturity **B**.
........... *Sympetrum fonscolombii*

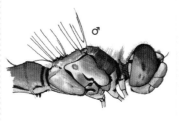

**NORTH:** JFMAMJJASOND
**SOUTH:** JFMAMJJASOND

**DISTRIBUTION:** a common resident in the Mediterranean, rarer further nor Undertakes long-range migrations en masse.

**HABITAT:** sunny, stagnant (sometimes brackish) waters in lowlands, includi ponds, lakes, gravel pits, coastal marshes and rice paddies.

**CONFUSION SPECIES:** *S. flaveolum*.

**6'.** Basal yellow spot on hindwings encompassing, at least in part, discoidal cell **A**.
Inferior surface of eyes yellow, brownish or reddish at maturity **B**.
............. *Sympetrum flaveolum*

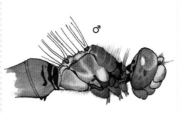

**NORTH:** JFMAMJJASOND
**SOUTH:** JFMAMJJASOND

**DISTRIBUTION:** found across the regio more sporadically in the west and co to higher altitudes in the south.

**HABITAT:** grassy ponds, marshes, flo meadows, peat bogs; in mountain ra and up to over 2,100m in the south o range.

**CONFUSION SPECIES:** *S. fonscolombii*.

## *Sympetrum depressiusculum* | Spotted Darter/Marshland Darter

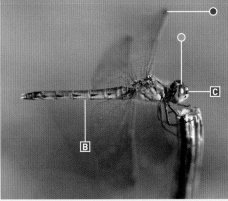

● Eyes touching at one point

● Pterostigmas pale red or yellow

● Legs entirely black

♂ abdomen: 20–23mm

♀ abdomen: 21–24mm

## *Sympetrum fonscolombii* | Red-veined Darter

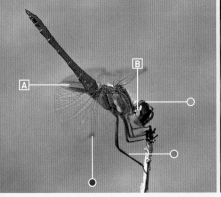

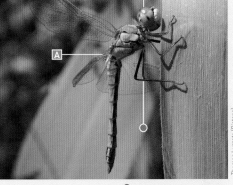

● Eyes touching at one point; inferior side of the eyes bluish-grey at maturity in males

● Pterostigmas pale yellow

● Legs striped with black and yellow

Thomas Luzzato (Biotope)

♂ abdomen: 22–29mm

♀ abdomen: 22–28mm

## *Sympetrum flaveolum* | Yellow-winged Darter

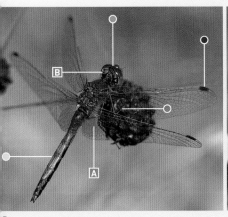

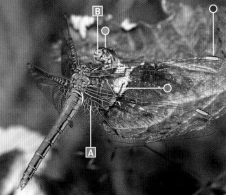

● Eyes touching at one point; inferior side of the eyes never bluish

● Pterostigmas red or yellow

● Legs striped with black and yellow

● Sides of the abdomen highlighted with black

♂ abdomen: 21–25mm

♀ abdomen: 19–27mm

**7.** Thoracic sutures usually almost completely devoid of black 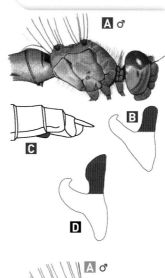.
♂ Outer branch of hamuli cylindrical and rounded, not widening towards end **B**.
♀ Vulvar plate appressed against abdomen, and barely visible in lateral view **C**.
.......... **Sympetrum meridionale**

**7'.** Thoracic sutures variable.
♂ Outer branch of hamuli spatula-shaped and widening towards end **D**.
♀ Vulvar plate partly or fully erect in relation to the abdomen **C** **C** ........ **8**

**8.** Horizontal black line at top of frons never bent downwards along eyes **B**.
♂ Sides of thorax with two yellow bands surrounding a median red band at maturity; all becoming darker with age **A**.
♀ Vulvar plate erect, at an oblique angle in relation to abdomen **C**.
.......... **Sympetrum striolatum**

**8'.** Horizontal black line at top of frons variable **B**.
♂ Sides of thorax yellow, brown or reddish brown, but without alternating red and yellow bands **A**.
♀ Vulvar plate erect, at right angles to abdomen **C**.
............. **Sympetrum vulgatum**
(two subspecies)

• Thoracic sutures clearly highlighted in black **A**. Legs black, with yellow or reddish-brown streaks. Horizontal black line at top of frons always bent downwards along eyes **E**.
Thorax becoming more or less dark reddish brown and abdomen becoming bright red. Robust appearance, measuring more than 23mm including abdomen and appendages.
.... **Sympetrum vulgatum vulgatum**

• Thoracic sutures very finely and incompletely highlighted in black **D**.
Legs yellowish brown with a few dark brown spots on femurs and tibiae. Horizontal black line at top of frons bent or not. Thorax remaining yellow or pale brown; abdomen yellowish brown, becoming brick red at a later stage. Slender appearance, measuring less than 24mm including abdomen and appendages.
.....**Sympetrum vulgatum ibericum**

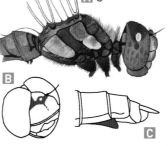

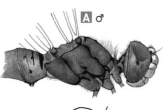

**NORTH:** JFMAMJJASOND
**SOUTH:** JFMAMJJASOND

**DISTRIBUTION:** common in places. A migratory species, spreading northwards thanks to warmer summers.

**HABITAT:** stagnant waters at low altitudes; individuals have been observed in the Alps, up to 3,000m.

**CONFUSION SPECIES:** S. striolatum, S. v. vulgatum and S. v. ibericum.

**NORTH:** JFMAMJJASOND
**SOUTH:** JFMAMJJASOND

**DISTRIBUTION:** common to very comm across the region, less so in the north-east.

**HABITAT:** sunny, stagnant and weakly flowing brackish waters. In the Mediterranean, abundant in autumn streams that are reduced to a few poc

**CONFUSION SPECIES:** S. vulgatum.

*Sympetrum*        *Sympetru*
*vulgatum vulgatum*   *vulgatum iber*

**NORTH:** JFMAMJJASOND
**SOUTH:** JFMAMJJASOND

**DISTRIBUTION:** S. v. vulgatum: common in central and eastern Europe, rarer further south and in the far north, ar occurs only as a vagrant in Britain. S. v. ibericum: northern Spain.

**HABITAT:** stagnant waters; up to 2,000m.

**CONFUSION SPECIES:** S. v. vulgatum with S. striolatum; S. v. ibericum with S. meridionale.

## Sympetrum meridionale | Southern Darter

● Eyes touching at one spot

● Legs mainly yellow-brown

● No large black patches at the end of the abdomen

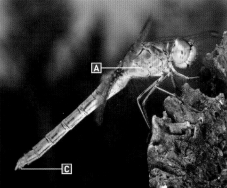

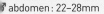
♂ abdomen: 22–28mm

♀ abdomen: 22–28mm

## Sympetrum striolatum | Common Darter

● Eyes touching at one spot

● Legs striped with black and yellow

♂ abdomen: 21–30mm

♀ abdomen: 20–30mm

## Sympetrum vulgatum vulgatum | Vagrant Darter/Moustached Darter

● Eyes touching at one spot

● Legs striped with black and yellow

Thomas Menut (Biotope)

♂ abdomen: 23–28mm

♀ abdomen: 24–28mm

## Genus *LEUCORRHINIA*

**1.** Cercoids white 🅐 🅐.
♂ Abdomen black with a well-developed
bluish pruinescence at base.
♀ Abdomen black with yellow spots.
. . . . . . . . . . . . . . . . . . . . . . . . . . . . . . . . **2**

**1'.** Cercoids black 🅐. Abdomen black with
red, brown or yellow spots in both sexes.
. . . . . . . . . . . . . . . . . . . . . . . . . . . . . . **3**

NORTH: JFMA**MJJAS**OND
SOUTH: JFM**AMJJA**SOND

DISTRIBUTION: relatively common
in the Baltic region and southern
Fennoscandia, but scattered
elsewhere. Has been expanding
west in the past few years.

HABITAT: stagnant eutrophic,
mesotrophic and oligotrophic waters
often in forested landscapes.

CONFUSION SPECIES: *L. albifrons*.

**2.** Abdomen strongly widened in segments
6–9 🅑. Two transverse cubito-anal veins
on hindwings, between discoidal cell and
base of wing.
♂ Abdomen covered in blue pruinescence
on segments (2)3–5 🅔. Pterostigmas
white above and blackish below 🅒.
♀ Lobes of vulvar plate acute and
reaching middle of ninth segment 🅓.
. . . . . . . . . . . . . . *Leucorrhinia caudalis*

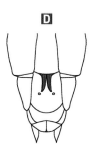

NORTH: JFMAM**JJAS**OND
SOUTH: JFMA**MJJA**SOND

DISTRIBUTION: locally common in the
Baltic region and Fennoscandia, but
scarce and local in the south and we[st]

HABITAT: peat bogs with *Sphagnum*
moss, acid peat ponds and former
lignite-mining areas; up to 1,150m.

CONFUSION SPECIES: *L. caudalis*, but in
*L. albifrons* the abdomen is only
slightly widened.

**2'.** Abdomen slightly widened or not widened at all in segments 6–9 🅑. Single
transverse cubito-anal vein on hindwings,
between discoidal cell and base of wing.
♂ Abdomen covered in blue pruinescence
on segments 3 and 4 🅔. Pterostigmas
black on both sides 🅒.
♀ Lobes of vulvar plate rounded and
reaching only one-fifth of length of ninth
segment 🅓.
. . . . . . . . . . . . . *Leucorrhinia albifrons*

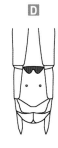

NORTH: JFMAM**JJAS**OND
SOUTH: JFM**AMJJ**ASOND

DISTRIBUTION: fairly common in the
Baltic region, but rare overall and
declining. Sporadic migrations likely
reinforce existing populations.

HABITAT: stagnant, fish-poor waters;
up to 2,000m in the south of Europe.

CONFUSION SPECIES: for ♀, with
*L. rubicunda*. Identification is possibl[e]
by examining the vulvar plate.

**3.** ♂ Clear dorsal spot of seventh abdominal
segment large, always remaining lemon
yellow; other spots becoming brown 🅑.
♀ Lobes of vulvar plate more or less
acute 🅒.
. . . . . . . . . . . . . *Leucorrhinia pectoralis*

## *Leucorrhinia caudalis* | Lilypad Whiteface/Dainty White-faced Darter

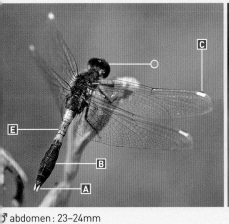

● Eyes touching at one spot

♂ abdomen: 23–24mm

♀ abdomen: 23–24mm

## *Leucorrhinia albifrons* | Dark Whiteface/Eastern White-faced Darter

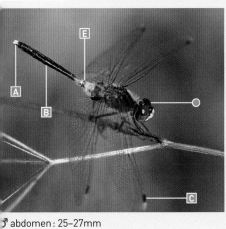

● Eyes touching at one spot

♂ abdomen: 25–27mm

♀ abdomen: 24–25mm

## *Leucorrhinia pectoralis* | Yellow-spotted Whiteface/Large White-faced Darter

● Eyes touching at one spot

♂ abdomen: 25–27mm

♀ abdomen: 23–26mm

**3'.** ♂ Clear spot on seventh abdominal segment red or yellow, and roughly same colour as spot on sixth segment (when it exists).
♀ Lobes of vulvar plate rounded at tip.
. . . . . . . . . . . . . . . . . . . . . . . . . . . . . . . . **4**

**4.** Basal black spots on forewings rather large, the anterior spot well marked; posterior spot encompassing cell that borders membranule **A**. Pterostigmas black in both sexes **B**.
♂ Clear spots on abdominal segments 4 and 5 small or absent **C**.
♀ Lobes of vulvar plate clearly distinct, and longer than wide **D**.
. . . . . . . . . . . . . . . . . *Leucorrhinia dubia*

**4'.** Basal black spots on forewings very small and punctate, the anterior spot absent or very small; posterior spot encompassing cell that borders membranule **A**. Pterostigmas red in ♂, black in ♀ **B**.
♂ Clear spots on abdominal segments 2–7 rather large and clearly visible **C**.
♀ Lobes of vulvar plate very short, only barely distinct, as wide as long, or wider than long **D**.
. . . . . . . . . . . *Leucorrhinia rubicunda*

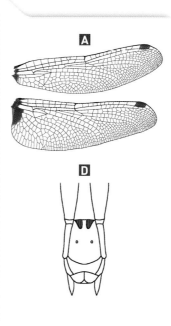

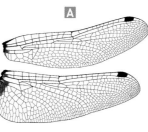

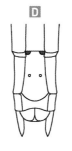

**North :** JFM**AMJJAS**OND
**South :** JFM**AMJJA**SOND

**Distribution:** the most common of the *Leucorrhinia* species, found across the region, although less so in the south.

**Habitat:** *Sphagnum* peat bogs, marshes and acidic, fish-poor ponds up to 2,300m.

**Confusion species:** *L. rubicunda*.

**North and South:** JFM**AMJJAS**OND

**Distribution:** northern Europe, reaching north of the Arctic Circle and as far south as Belgium; very rare and declining.

**Habitat:** stagnant, fish-poor, generally acidic waters at low altitudes (marshes, peat lakes and ponds, *Sphagnum* peat bogs, mesotrophic grassy ponds).

**Confusion species:** for ♂, with *L. dubia*; for ♀, with *L. pectoralis*.

## Leucorrhinia dubia | Small Whiteface/White-faced Darter

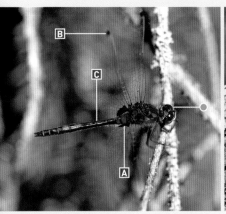

● Eyes touching at one spot

Thomas Roussel (Biotope)

♂ abdomen: 24–28mm

♀ abdomen: 21–25mm

## Leucorrhinia rubicunda | Ruby Whiteface

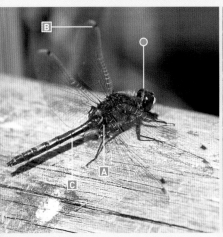

● Eyes touching at one spot

♂ abdomen: 26–28mm

♀ abdomen: 23–27mm

Banded Demoiselle, *Calopteryx splendens*

# Field Guide to the Exuviae and Larvae (Final Instar) of Dragonflies

## INTRODUCTION

The quality of the aquatic environment in which dragonfly larvae develop is often essential to the survival of the species.

The study of dragonfly larvae allows us to gain a good understanding of the quality of their habitats, to examine the suite of species linked to a particular habitat and to evaluate the density of populations. In addition, the collection of exuviae is a non-invasive sampling method for Odonata, because no living individuals are removed from the environment – exuviae are just the dried outer envelopes of young dragonflies, abandoned during emergence. Another advantage of collecting exuviae, especially those of large species, is that they are fairly easy to find in the riparian vegetation bordering water features and rivers.

Collection and identification of exuviae is recommended to confirm the presence of species on a site and to understand certain aspects of their biology. Indeed, the observation of adults is not sufficient to quantify dragonfly populations. Many species of dragonfly are territorial, with males excluding congeners from their territory. As a result, a large number of males may stay away from breeding sites. Some species rarely come to breeding grounds because of their secretive habits or vagrant behaviour. The most striking example of these 'hidden' species is undoubtedly *Epitheca bimaculata*. On certain sites the adults are only rarely observed, yet thousands of the species' exuviae may be collected during the emergence period!

In addition, collecting a large number of exuviae from a site makes it possible to determine the sex ratio (proportion of females compared to males) of the various species monitored on that site.

When identifying species, entomologists must remain vigilant and rigorous, and in particular they must examine the state of several identification criteria for each specimen. Use of a binocular microscope is often recommended. The key presented in this guide, which above all aims to be a tool requiring neither microscopic preparation nor dissection, proposes a simplified method for the identification of exuviae. It permits the identification of all families, most genera and many species. In some cases, the use of a ×10 magnifying glass and a ruler is required. Those who would like to improve their knowledge in this field can refer to the works of Heidemann & Seidenbusch (2002), Gerken & Sternberg (1999), Doucet (2016) and Brochard et al. (2012).

Take care with dry exuviae, as they are very fragile to handle. They are also often set in a position that does not allow for proper study. For certain species, it may therefore sometimes be necessary to prepare the exuvia by softening it for a few hours in a moist atmosphere, so that it can be spread out completely and the legs moved.

*Note: In the following dichotomous keys, the length given for the exuviae is only that of the body (head + thorax + abdomen); the legs are excluded from the measurements. Within the same species, the measurements of different exuviae can vary by more than 20 per cent. As a result, some exuviae may appear abnormally small compared to the average across all samples collected.*

# Key distinctions between the exuviae of
# Zygoptera and Anisoptera

Exuviae slender and elongated; maximum width of abdomen 2–3mm **A**.
Caudal lamellae present at end of abdomen* **B**.
. . . . . . . . . . . . . . . Zygoptera (Table 1, p. 108)

The lamellae are sometimes absent on the exuviae; in this case, the first character should be trusted.

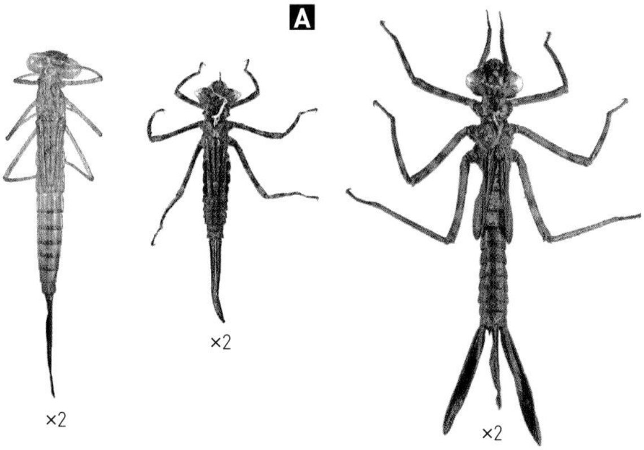

**A**

×2

×2

×2

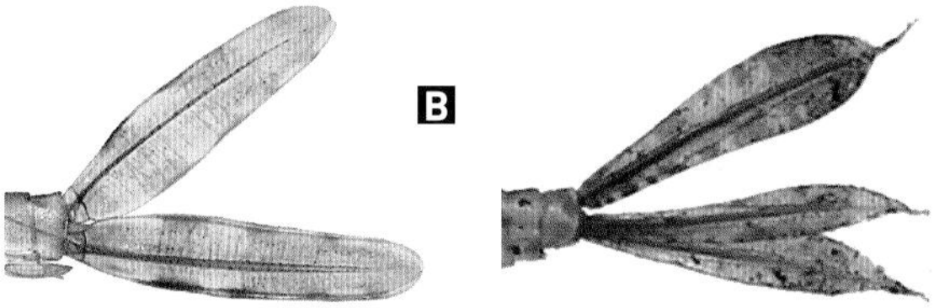

**B**

. Exuviae rather robust; minimum width of the abdomen 4mm **C**.
Anal pyramid present at end of abdomen **D**.
. . . . . . . . . . . . . Anisoptera (Table A, p. 114)

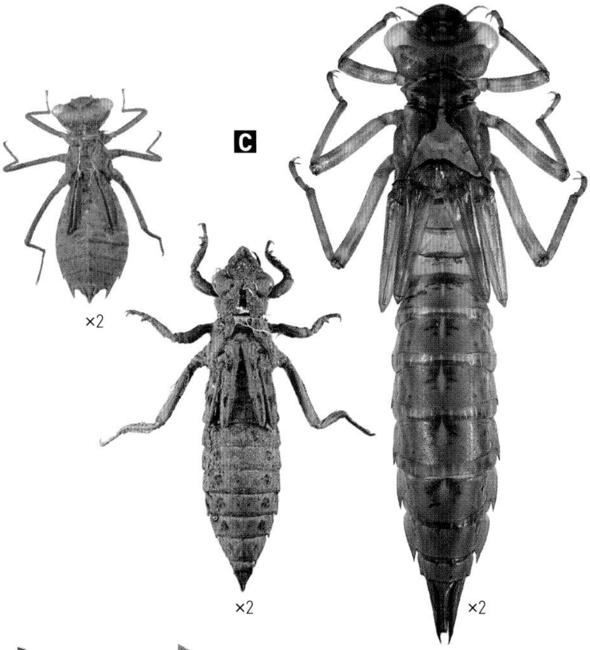

**C**

×2

×2

×2

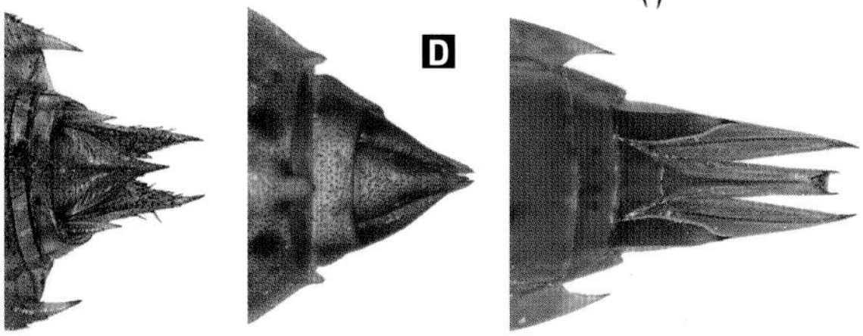

**D**

Table 1

# Key to the exuviae of Zygoptera

**1.** Mentum with a diamond-shaped hole **A**.
First article of antenna longer than rest of antenna **B**.
. . . . . . . . . . . . . . . . Calopterygidae (Table 2, p. 109)

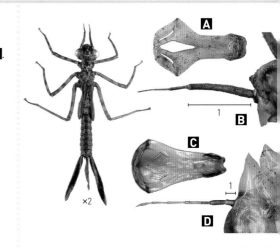

**1'.** Mentum without hole **C**.
First article of antenna short **D**.
. . . . . . . . . . . . . . . . . . . . . . . . . . . . . . . . . . . . . . . .2

**2.** Labial palps split into two parts by a deep notch*
(examine mask from front) **E**.
. . . . . . . . . . . . . . . . . . . . . . . Lestidae (Table 3, p. 109)

deep notch

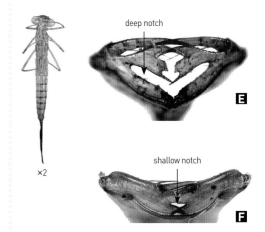

**2'.** Labial palps without deep notch* (examine mask
from front) **F**.
. . . . . . . . . . . . . . . . . . . . . . . . . . . . . . . . . . . . . . . .3

\* This character can be difficult to assess when using
a simple magnifying glass (dirty exuviae, antennae
folded over the palps ...). If in doubt, bring the exuviae
to the lab and examine the palps under a binocular
microscope.

shallow notch

**3.** Caudal lamellae terminating in a long filament.
Median trachea protruding **G**.
. . . . . . . . . . . . Platycnemididae, Genus *Platycnemis*

⚠ Note: The exuviae of the three species present in
the area considered in this book are not significantly
different from one another.

median trachea protruding

sectional
view
**G**

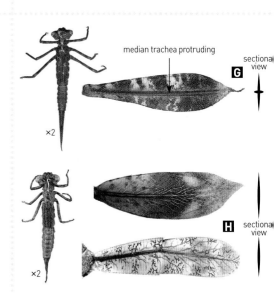

**3'.** Caudal lamellae sometimes with a pointed tip,
never terminating in a long filament.
Median trachea of caudal lamellae at least partly
visible, but not protruding **H**.
. . . . . . . . . . . . . . Coenagrionidae (Table 6, p. 111)

**H** sectional
view

Table 2

# Family Calopterygidae (Genus *Calopteryx*)

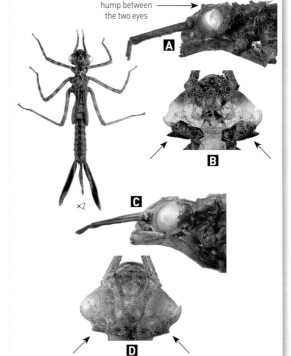

Very pronounced hump between the two eyes **A**.
Occiput with well-developed and pointed posterior expansions **B**.
. . . . . . . . . . . . . . . . . . . . . . . . . . . . . . . . . *Calopteryx virgo*

hump between the two eyes

. No hump or a very slight hump between the two eyes **C**.
Occiput with reduced and blunt posterior expansions **D**.
. . . . . . . . . . . . . . . . . . . . . Other *Calopteryx* species
(*Calopteryx splendens*, *Calopteryx haemorrhoidalis* or *Calopteryx xanthostoma*)

⚠ Note: The larvae and exuviae of these species cannot be distinguished from one another in the field.

×2

---

Table 3

# Family Lestidae

. Mentum stalked **A**.
. . . . . . . . . . . . . . . . Genus *Lestes* (Table 4, p. 110)

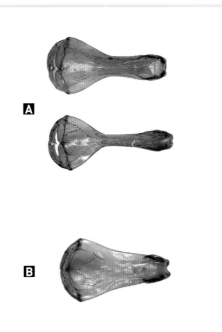

'. Mentum unstalked **B**.
. . . . . . . . . . . Genera *Chalcolestes* and *Sympecma*
(Table 5, p. 111)

Table 4

# Genus *Lestes*

**1.** Stalk broad (maximum width ≈ 3× minimum width) **A**.
. . . . . . . . . . . . . . . . . . . . . . . . . . . . . *Lestes macrostigma*

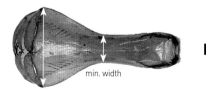

**A**

min. width

max. width

**1'.** Stalk narrow (maximum width ⩾4× minimum width) **B**.
. . . . . . . . . . . . . . . . . . . . . . . . . . . . . . . . . . . . . . . .2

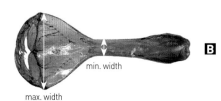

**B**

min. width

max. width

**2.** Caudal lamellae acute at tip* **C**.
. . . . . . . . . . . . . . . .*Lestes dryas* and *Lestes barbarus*

**C**

**2'.** Caudal lamellae rounded at tip* **D**.
. . . . . . . . . . . . . . . . . . . *Lestes sponsa* and *Lestes virens*

*  The lamellae are sometimes folded at the tip.
   To examine the lamellae it is therefore important
   to wet them and spread them correctly.

**D**

# Genera *Chalcolestes* and *Sympecma*

Last tooth of comb of labial palp large and separated from others by a groove* **A**.
. . . . . . . . . . . . . . . . . . . . . . . . . . . Genus *Sympecma*

⚠ Note: The two species in this genus are very difficult to distinguish from each other, but *Sympecma paedisca* is very localised.

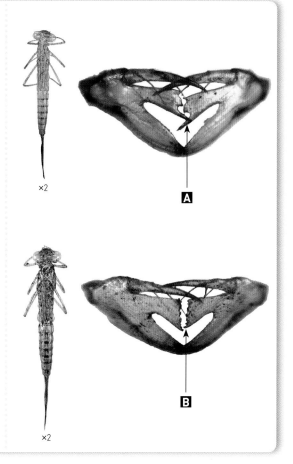

. Last tooth of comb of labial palp slightly longer than others. No groove* **B**.
. . . . . . . . . . . . . . . . . . . . . . . . . Genus *Chalcolestes*

⚠ *Chalcolestes viridis* and *C. parvidens* overlap in part of our region; separating these species using their exuviae is very difficult.

\* This character can be difficult to assess when using a simple magnifying glass (dirty exuviae, antennae folded over the palps …). If in doubt, bring the exuviae to the lab and examine the palps under a binocular microscope.

---

# Family **Coenagrionidae**

entifying the exuviae and larvae of this family is extremely difficult and falls largely outside the scope of this ork. However, some information is provided here on e key taxa.

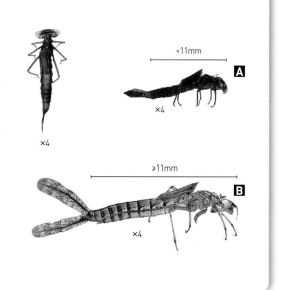

Body length (without procts) <11mm **A**.
. . . . . . . . . . . . . . . . . . . . . . . . . . . *Nehalennia speciosa*

⚠ Note: A very rare species across most of the region.

. Body length (without procts) ≥11mm **B**.
. . . . . . . . . . . . . . . . . . . . . . . . . . . . . . . . . . .2

Table 6
... continued

# Family Coenagrionidae ... continued

**2.** Small spines present on cuticle of ventral part of first abdominal segment **C**.
............Genus *Erythromma* (Table 7, p. 113)

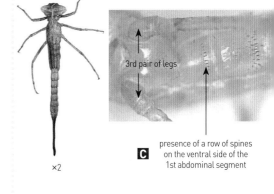

3rd pair of legs

**C** presence of a row of spines on the ventral side of the 1st abdominal segment

×2

**2'.** Absence of small spines on cuticle of ventral part of first abdominal segment **D**.
.........................................3

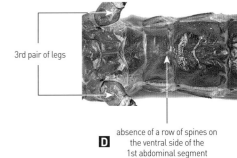

3rd pair of legs

**D** absence of a row of spines on the ventral side of the 1st abdominal segment

**3.** On the head, in dorsal view, occiput forming a simple curve with back of eye **E**.
..... Genera *Enallagma*, *Ischnura* and *Coenagrion*

⚠ Note: For specific identification, use a binocular microscope and consult the works cited in the bibliography on p. 147.

**3'.** On the head, in dorsal view, occiput forming a marked angle with back of eye **F H**.
.........................................4

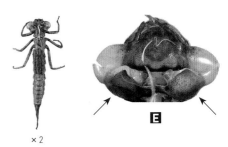

× 2

**E**

# Family Coenagrionidae ... continued

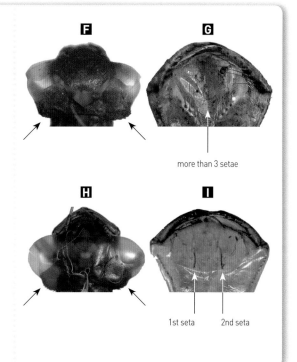

Occiput with a very marked angle (more so than in *Ceriagrion tenellum*) **F**.
More than three long setae inside mask (central part)* **G**.
. . . . . . . . . . . . . . . . . . . . . . . . . . . Genus *Pyrrhosoma*
(a single species: *Pyrrhosoma nymphula*)

more than 3 setae

Occiput with a more rounded angle (less acute than in *Pyrrhosoma nymphula*) **H**.
Two (sometimes three) long setae inside mask (central part)* **I**.
. . . . . . . . . . . . . . . . . . . . . . . . . . .Genus *Ceriagrion*
(a single species: *Ceriagrion tenellum*)

The use of a binocular microscope is recommended to examine this character.

1st seta      2nd seta

# Genus *Erythromma*

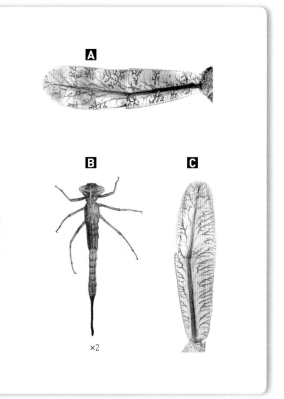

Length of exuviae ≥24mm. Body length (without procts) 16–22mm. Procts strongly pigmented; three dark transverse bands at end of procts **A**.
. . . . . . . . . . . . . . . . . . . . . . . . . . . *Erythromma najas*

Length of exuviae ≤22mm **B**. Body length (without procts) 12–16mm. Procts only weakly pigmented or not pigmented at all **C**.
. . . . . . . . . . . . . . . . . . . . . . . . . . . *Erythromma lindenii*
or *Erythromma viridulum*

⚠ Note: For specific identification, use a binocular microscope and consult the works cited in the bibliography on p. 147.

×2

# Key to the exuviae of Anisoptera

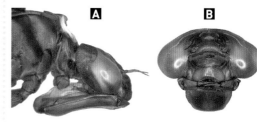

**1.** Mask flat **A**.
   Labial palps do not cover clypeus or labrum **B**.
   . . . . . . . . . . . . . . . . . . . . . . . . . . . . . . . . . . . . . . . . . . . .2

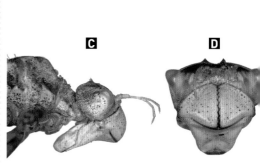

**1'.** Mask spoon-shaped **C**.
   Labial palps cover clypeus and labrum **D**.
   . . . . . . . . . . . . . . . . . . . . . . . . . . . . . . . . . . . . . . . . . .3

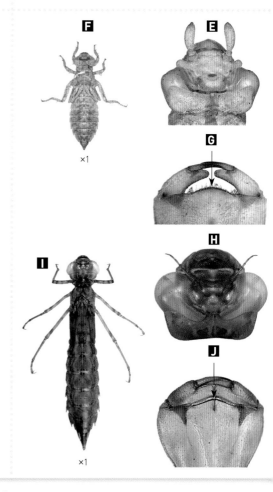

**2.** Exuviae compact, with abdomen more
   or less flattened dorsoventrally **F**.
   Antennae club-shaped* **E**.
   No groove at tip of mask **G**.
   . . . . . . . . . . . . . . . . . . Gomphidae (Table B, p. 116)

**2'.** Exuviae more elongated; abdomen not flattened
   dorsoventrally **I**.
   Antennae filiform* **H**.
   Groove at tip of mask **J**.
   . . . . . . . . . . . . . . . . . . Aeshnidae (Table C, p. 117)

   * Most of the time the antennae are appressed along
   the eyes and the mask.

Mask with strongly indented lip palps, bearing irregular teeth **K**.
Total length of exuviae ⩾35mm **L**.
. . . . . . . . . Cordulegastridae, Genus *Cordulegaster*
(Table J, p. 129)

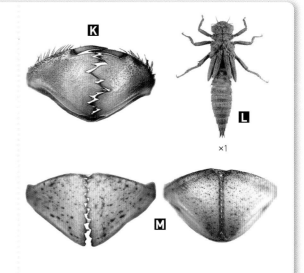

×1

Mask with lip palps bearing regular teeth, or toothless **M**.
Total length of exuviae ⩽35mm.
. . . . . . . . . . . . . . . . . . . . . . . . . . . . . . . . . . . . . .4

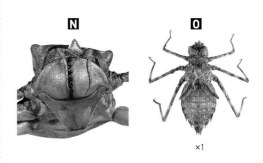

Distinct large conical protrusion on frons **N**.
Total length of exuviae 30–34mm **O**.
. . . . . . . . . . . . . . . . . . . . . . . . . . . . . . Macromiidae
(a single species: *Macromia splendens*)

No conical protrusion on frons. Total length of exuviae <33mm.
. . . . . . . . . . . . . . . . . . . . . . . . . . . . . . . . . . . . . .5

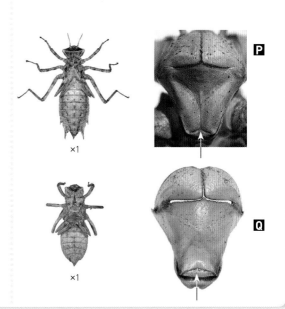

×1

Mentum with a small groove (sometimes weakly marked) at base **P**.
. . . . . . . . . . . . . . Corduliidae and *Oxygastra curtisii*
(Table D, p. 119)

Mentum with a groove at base **Q**.
. . . . . . . . . . . . . . . . . . Libellulidae (Table E, p. 121)

×1

×1

# Family Gomphidae

**1.** Lobe of mentum between the two labial palps well developed: height of lobe (h lobe)/base lobe >0.2 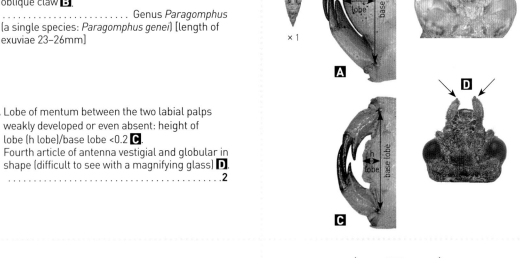. Fourth article of antenna very small compared to third article, elongated, in the shape of a bent or oblique claw **B**.
. . . . . . . . . . . . . . . . . . . . . . . . Genus *Paragomphus* (a single species: *Paragomphus genei*) [length of exuviae 23–26mm]

**1'.** Lobe of mentum between the two labial palps weakly developed or even absent: height of lobe (h lobe)/base lobe <0.2 **C**. Fourth article of antenna vestigial and globular in shape (difficult to see with a magnifying glass) **D**.
. . . . . . . . . . . . . . . . . . . . . . . . . . . . . . . . . . .2

**2.** Prominent part of labial palp (positioned against mobile hook) pointed* **E**.
. . . . . . . . . . . . . . . . . . . . . . . . . . . . . . . . . . .3

**2'.** Prominent part of labial palp (positioned against mobile hook) rounded* **F**.
. . . . . . . . . . . . . . . . . . . . . . . . . . . . . . . . . .4

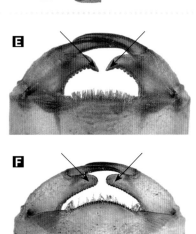

\* This character can be difficult to assess when using a simple magnifying glass (dirty exuviae). If in doubt, bring the exuviae to the lab and examine the palps under a binocular microscope.

**3.** Spines present (at least on segments 5–8) **G**.
. . . . . . . . . . . . . . . . . . . . . . . . . . . . . Genus *Lindenia* (a single species: *Lindenia tetraphylla*) [length of exuviae 28–45mm]

⚠ Note: This species has so far only been reported from Corsica.

**3'.** Spines absent **H**.
. . . . . . . . . . . . . Genus *Gomphus* (Table F, p. 123)

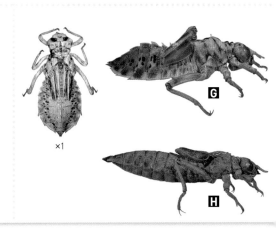

# Family Gomphidae ... continued

Length of segment 9 (S9)/length of segment 10 (S10) ⩾1.6 **I**.
In frontal view, lateral lobes of frons point downwards* **J**.
Length of exuviae ⩾27mm.
........................Genus *Ophiogomphus*
(a single species: *Ophiogomphus cecilia*) [length of exuviae 27–32mm] **K**.

. Length of segment 9 (S9)/length of segment 10 (S10) <1.6 **L**.
In frontal view, lateral lobes of frons spread obliquely outwards and extend beyond edge of eyes* **M**.
Length of exuviae ⩽26mm.
......................Genus *Onychogomphus*
(Table G, p. 125) **N**.

* This character can be difficult to assess when using a simple magnifying glass (dirty exuviae). If in doubt, bring the exuviae to the lab and examine the palps under a binocular microscope.

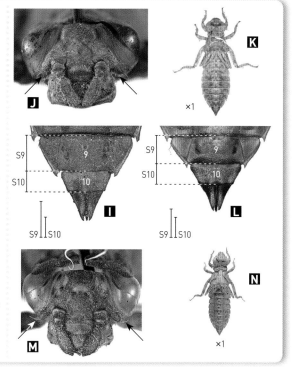

# Family Aeshnidae

Table C

Occiput angular **A**.
Lateral spines from the fifth segment onwards **B**.
.................................Genus *Boyeria*
(a single species: *Boyeria irene*) [length of exuviae 32–44mm]

. Occiput rounded **C**.
No lateral spines from fifth segment onwards.
.........................................2

In dorsal view, length of occiput is greater than diameter of eye **D**.
...........................Genus *Brachytron*
(a single species: *Brachytron pratense*) [length of exuviae 35–40mm]

. In dorsal view, length of occiput is smaller than diameter of eye **E**.
........................................3

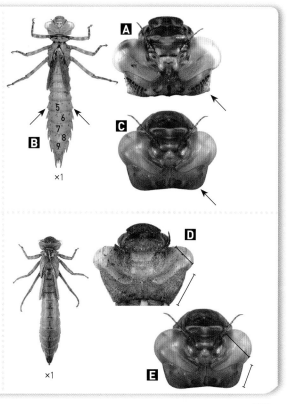

Table C
... continued

# Family Aeshnidae ... continued

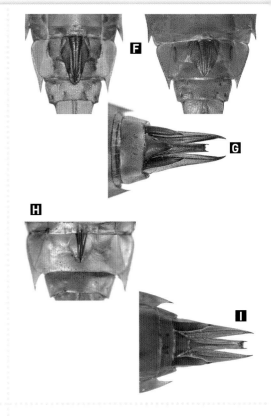

**3. ♀** Ovipositor almost reaching, or extending beyond, ninth abdominal segment **F**.
**♂** Anal pyramid with a thickened expansion, longer than wide, on top and at base of epiproctus, ending in an obtuse point **G**.
. . . . . . . . . . . . . . . . .Genus *Aeshna* (Table H, p. 126)

**3'. ♀** Ovipositor not reaching rear edge of ninth abdominal segment **H**.
**♂** Anal pyramid with a thickened expansion (barely visible in *Hemianax*), not as long as wide, on top and at base of epiproct, ending in a wide and linear or slightly concave tip **I**.
. . . . . . . . . . . . . . . . . . . . . . . . . . . . . . . . . . . . . . .4

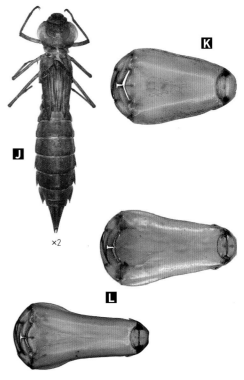

**4.** Length of exuviae 42–46mm **J**.
Mask stout **K**.
. . . . . . . . . . . . . . . . . . . . . . . . . . . . Genus *Hemianax*
(a single species: *Hemianax ephippiger*)

×2

**4'.** Length of exuviae >46mm.
Mask fairly elongated **L**.
. . . . . . . . . . . . . . . . . .Genus *Anax* (Table I, p. 128)

# Family Corduliidae and *Oxygastra curtisii*

. Head carrying two pointed tubercles on vertex **A**. Lateral spines of segment 9 reaching or exceeding end of anal pyramid **B**. Length of exuviae ⩾27mm.
. . . . . . . . . . . . . . . . . . . . . . . . . . . . Genus *Epitheca* (a single species: *Epitheca bimaculata*) [length of exuviae 27–32mm]

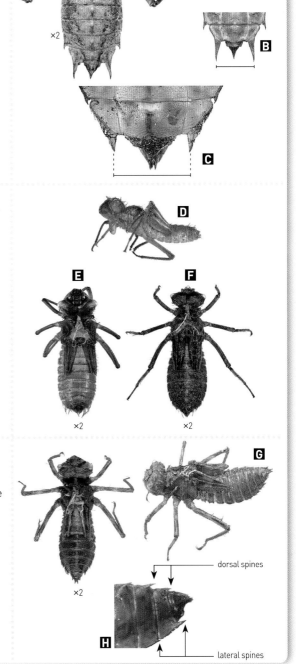

×2

**A**

**B**

. Head not carrying true tubercles on vertex, but sometimes bearing small humps. Lateral spines of segment 9 not reaching end of anal pyramid **C**. Length of exuviae ⩽26mm.
. . . . . . . . . . . . . . . . . . . . . . . . . . . . . . . . . . . . . .2

**C**

. Absence of lateral and dorsal spines on abdomen **D**.
. . . . . . . . . . . . . . . . . . . . . . . . .Genus *Somatochlora* of the *Somatochlora arctica* **E** and *Somatochlora alpestris* **F** group [exuviae hairy; length of exuviae 17–22mm long]

⚠ Note: For specific identification, use a binocular microscope and consult the works cited in the bibliography on p. 147.

**D**

**E**        **F**

. Spines present on abdomen, at least laterally on segments 8 and 9.
. . . . . . . . . . . . . . . . . . . . . . . . . . . . . . . . . . . . . .3

×2        ×2

. Dorsal spines replaced by tufts of clustered bristles (exuviae appearing hairy) **G**.
. . . . . . . . . . . . . . . . . . . . . . . . . . . Genus *Oxygastra* (a single species: *Oxygastra curtisii*) [length of exuviae 19–22mm]

**G**

. True dorsal and lateral spines present on abdomen (exuviae appearing glabrous) **H**.
. . . . . . . . . . . . . . . . . . . . . . . . . . . . . . . . . . . . . .4

×2

dorsal spines

lateral spines

**H**

**4.** Spine on segment 9 only slightly developed (or
absent) **I**.
. . . . . . . . . . . . . . . . . . . . . . . . . . . . Genus *Cordulia*
(a single species: *Cordulia aenea*) [length of exuviae
21–26mm]

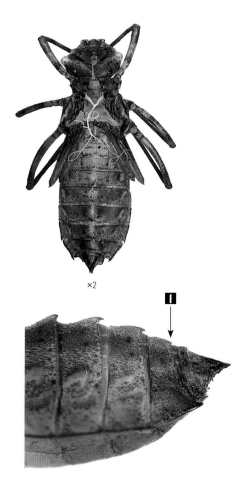

×2

**4'.** Spine on segment 9 well developed **J**.
. . . . . . . . . . . . . . . . . . . Genus *Somatochlora* of the
*Somatochlora metallica*, *Somatochlora flavomaculata*
and *Somatochlora meridionalis* group (Table K, p. 130)

# Family Libellulidae

Lateral edges of head distinctly parallel (top view) **A**.
Diameter of eye small relative to head (less than half the length of head) **B**.
. . . . . . . . . . . . . . . . . . . . . . . . . . . . . . . . . . . . . . . .2

. Lateral edges of head clearly converging towards rear (top view) **C**.
Diameter of eye larger relative to head (more than half the length of head) **D**.
. . . . . . . . . . . . . . . . . . . . . . . . . . . . . . . . . . . . .3

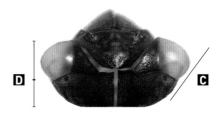

Dorsal spine on segment 7 **E**.
. . . . . . . . . . . . . . . Genus *Libellula* (Table L, p. 131)

. No dorsal spine on segment 7 **F**.
. . . . . . . . . . . . .Genus *Orthetrum* (Table M, p. 131)

**3.** Dorsal spine on segment 9 well developed* 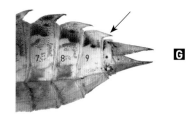 .
No lateral spine on segment 7 .
. . . . . . . . . . . . Genera *Trithemis* and *Brachythemis*
(Table N, p. 133)

\* Any beginner to identifying exuviae may find it
difficult to recognise this character. If in doubt,
refer to a specialist for help and confirmation.

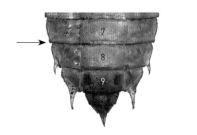

**3'.** Not as above . . . . . . . . . . . . . . . . . . . . . . . . . . . . . .4

**4.** Eyes clearly prominent, almost conical and acute on
each side of head* 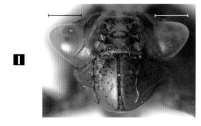.
. . . . . . . . . . . . Genus *Leucorrhinia* (Table O, p. 134)

\* Any beginner to identifying exuviae may find it
difficult to recognise this character. If in doubt,
refer to a specialist for help and confirmation.

**4'.** Eyes only slightly prominent, or not prominent at all,
and almost globular on each side of head .
. . . . . . . . Genera *Crocothemis*, *Selysiothemis* and
*Sympetrum* (Table P, p. 135)

# Genus *Gomphus*

Mentum longer than wide: length (L)/width (W) ≈ 1.15 **A**.
In ventral view, ninth abdominal segment slightly longer than its largest width **B**.
Length of exuviae ⩾32mm.
. . . . . . . . . . . . . . . . . . . . . . . . . . . . *Stylurus flavipes*
[length of exuviae 32–35mm]

. Mentum about as long as its maximum width **C**.
In ventral view, ninth abdominal segment not as long as wide **D**.
Length of exuvia <32mm. . . . . . . . . . . . . . . . . . . . . .**2**

. In ventral view, 10th abdominal segment almost square: length ≈ width **E**.
. . . . . . . . . . . . . . . . . . . . . . . . *Gomphus pulchellus*
[length of exuviae 25–30mm]

'. In ventral view, 10th abdominal segment rather rectangular **F**.
. . . . . . . . . . . . . . . . . . . . . . . . . . . . . . . . . . . . .**3**

# Genus *Gomphus* ... continued

**3.** Abdominal segments 6–9 with lateral spines* 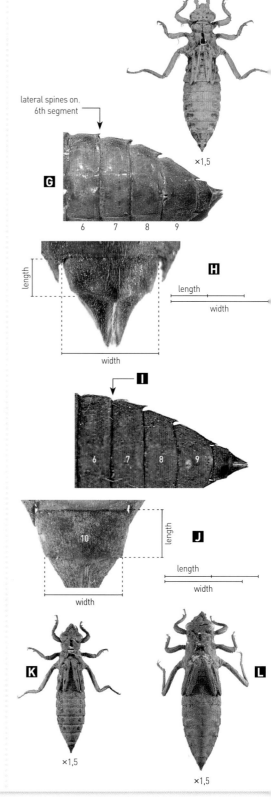.
In ventral view, 10th abdominal segment more than
twice as wide as long **H**.
. . . . . . . . . . . . . .*Gomphus vulgatissimus*
[length of exuviae 24–32mm]

\* Exuviae often need to be cleaned for this character
to be clear.

lateral spines on
6th segment

×1,5

**G**

6  7  8  9

length

**H**

length
width

width

**I**

6  7  8  9

**3'.** Abdominal segment 6 without lateral spines **I**.
In ventral view, 10th abdominal segment less than
twice as wide as long **J**.
. . . . *Gomphus graslinii* **K** or *Gomphus simillimus* **L**
[length of exuviae 25–30mm]

⚠ Note: It is particularly difficult to separate the
exuviae of these two species. To do so, it is necessary
to use a binocular microscope and consult the works
cited in the bibliography on p. 147.

10

length

**J**

length
width

width

**K**

**L**

×1,5

×1,5

# Genus *Onychogomphus*

Lateral spines present only on abdominal segments 8 and 9 **A**.
Dorsal spines generally well marked, rarely reduced to tubercles (regional variation) **B**.
. . . . . . . . . . . . . . . . . . . . . .*Onychogomphus uncatus*
[length of exuviae 22–26mm]

⚠ Note: In some populations of *Onychogomphus uncatus* the dorsal spines are reduced to tubercles, as in *Onychogomphus forcipatus*. Confusion is therefore possible, and in this case only the absence or presence of lateral spines on the seventh abdominal segment remains a usable character for identification.

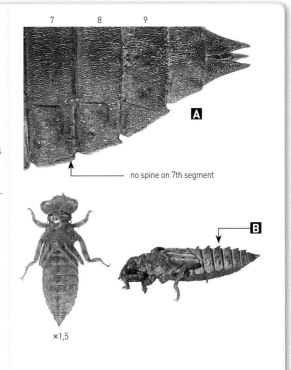

no spine on 7th segment

×1,5

lateral spines on 7th segment

Lateral spines present on abdominal segments 6/7–9 **C**.
Dorsal spines not very marked, often reduced to tubercles **D**.
. . . . . . . . . . . . . . . . . . . . *Onychogomphus forcipatus*
[length of exuviae 22–26mm]

⚠ Note: This taxon includes two subspecies.
• *Onychogomphus forcipatus forcipatus*: generally, lateral spines present on segment 6 (absent, vestigial or present on one side only in 6 per cent of cases).
• *Onychogomphus forcipatus unguiculatus*: generally, no lateral spines on segment 6 (present, vestigial or absent on one side only in 7 per cent of cases).

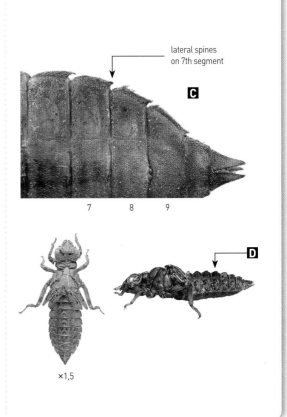

×1,5

# Genus *Aeshna*

The exuviae of the 8 species in the genus *Aeshna* are extremely difficult to identify in the field

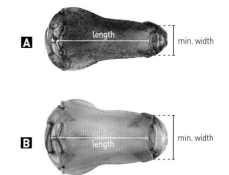

**1.** Mask stalked (length (L)/minimum width (W) >2.6) **A**.
. . . . . . . . . . . . . . . . . . . . . . . . . . . . . . . . . . . . . . .2

**1'.** Mask unstalked (length (L)/minimum width (W) <2.6) **B**.
. . . . . . . . . . . . . . . . . . . . . . . . . . . . . . . . . . . . . .4

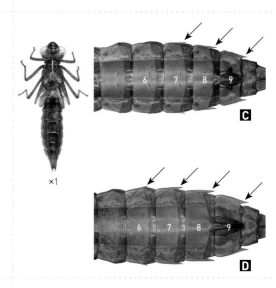

**2.** Lateral spines only on segments 7–9 (sometimes very small outlines on segment 6) **C**.
. . . . . . . . . . . . . . . . . . . . . . . . . . . . . *Aeshna caerulea*
[length of exuviae 34–42mm]

⚠ Note: Species present only in the Alps, above 1,000m in Haute-Savoie and Switzerland.

**2'.** Lateral spines on segments 6–9 (those of segment 6 clearly visible) **D**.
. . . . . . . . . . . . . . . . . . . . . . . . . . . . . . . . . . . . . .3

×1

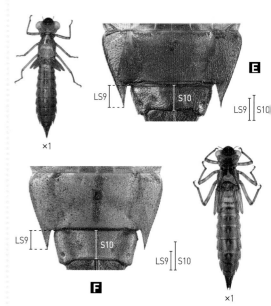

**3.** Length of lateral spines on segment 9 (LS9) more than three-quarters the length of segment 10 (S10) **E**.
Length of exuviae 30–38mm.
. . . . . . . . . . . . . . . . . . . . . . . . . . . . . *Aeshna mixta*

×1

**3'.** Length of lateral spines on segment 9 (LS9) less than three-quarters the length of segment 10 **F**.
Length of exuviae 38–48mm.
. . . . . . . . . . . . . . . . . . . . . . . . . . . . *Aeshna cyanea*

×1

# Genus *Aeshna* ... continued

Length of cerci more than half the length of paraprocts (≥0.6)* **G**.
..................................................**5**

, Length cerci/length of paraprocts <0.6* **H**.
.........................................**6**

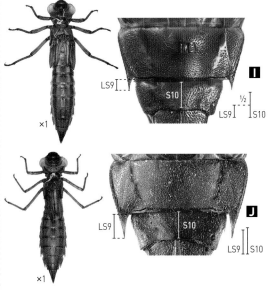

Lateral spines of segment 9 (LS9) less than half the
length of segment 10 (S10) **I**.
Lateral spines only on segments 7–9.
.............................*Aeshna subarctica*
[length of exuviae 37–43mm]

, Lateral spines of segment 9 (LS9) almost the same
length as segment 10 (S10) **J**.
Lateral spines present on segments 6–9 (rarely
absent on segment 6).
...............................*Aeshna isoceles*
[length of exuviae 38–44mm]

Length of lateral spines of segment 9 (LS9)/length
of segment 10 (S10) >0.6* **K**.
Length of exuviae 29–39mm.
.................................*Aeshna affinis*

, Length of lateral spines of segment 9 (LS9)/length
of segment 10 <0.6* **L**.
Length of exuviae 37–46mm.
.........................................**7**

This character can be difficult to assess when using
a simple magnifying glass (dirty exuviae). If in doubt,
bring the exuviae to the lab and examine the palps
under a binocular microscope.

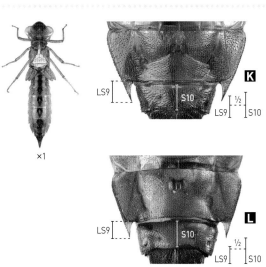

**7.** In most cases, two light spots present on darker
surfaces at back of occiput **M**.
. . . . . . . . . . . . . . . . . . . . . . . . . . . . . . . *Aeshna grandis*
[length of exuviae 40–46mm]

×1

**7'.** Occiput a homogeneous colour (no light spot) **N**.
. . . . . . . . . . . . . . . . . . . . . . . . . . . . . *Aeshna juncea*
[length of exuviae 37–45mm]

×1

Table I

# Genus *Anax*

**1.** Distal width of expansion (dist. W. exp.)/height of
expansion (H. exp) <0.8 **A**.
. . . . . . . . . . . . . . . . . . . . . . . . . . . . . *Anax imperator*
[length of exuviae 45–59mm]

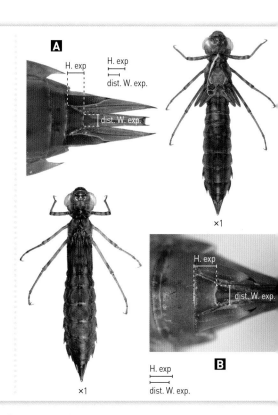

×1

**1'.** Distal width of expansion (dist. W. exp.)/height of
expansion (H. exp) >0.9 **B**.
. . . . . . . . . . . . . . . . . . . . . . . . . . . *Anax parthenope*
[length of exuviae 48–54mm]

×1

# Genus *Cordulegaster*

Lateral spines on segments 8 and 9 **A**.
..........................*Cordulegaster boltonii*
[length of exuviae 35–47mm]

×1

No lateral spines on segments 8 or 9 **B**.
.........................*Cordulegaster bidentata*
[length of exuviae 34–45mm]

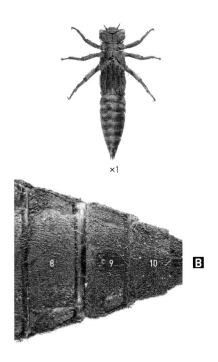

×1

# Genus *Somatochlora* (excluding *S. arctica* and *S. alpestris*)

**1.** At best, lateral spines on segment 9 (LS9) only
slightly longer than those on segment 8 (LS8) **A**.
. . . . . . . . . . . . . . . . . . . . . . . *Somatochlora metallica*
[length of exuviae 22–26mm]

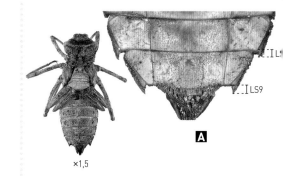

×1,5

**1'.** Lateral spines on segment 9 (LS9) clearly longer
than those on segment 8 (LS8) **B**.
. . . . . . . . . . . . . . . . . . . . . . . . . . . . . . . . . . . . .**2**

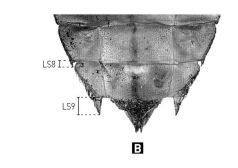

**2.** Dorsal spines present on segments 3–9* **C**.
. . . . . . . . . . . . . . . . . . . . . . *Somatochlora meridionalis*
[length of exuviae 20–24mm]

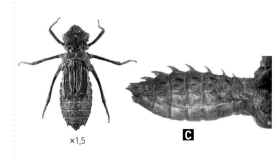

×1,5

**2'.** Dorsal spines present on segments (4)5–9* **D**.
. . . . . . . . . . . . . . . . . . . . .*Somatochlora flavomaculata*
[length of exuviae 19–22mm]

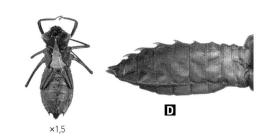

×1,5

\* It may be necessary to move the wing sheaths
in order to observe the first dorsal spines.

# Genus *Libellula*

Dorsal spine on segment 9 well developed **A**.
. . . . . . . . . . . . . . . . . . . . . . . . . . . . . . . . *Libellula fulva*
[length of exuviae 20–25mm]

No dorsal spine on segment 9 **B**.
. . . . . . . . . . . . . . . . . . . . . . . . . . . . . . . . . . .2

Teeth of labial palps distinctly marked **C**.
Cerci less than half the length of paraprocts **D**.
. . . . . . . . . . . . . . . . . . . . . . . . . . . *Libellula depressa*
[length of exuviae 21–26mm] **E**

Teeth of labial palps only slightly marked **F**.
Cerci more than half the length of paraprocts **G**.
. . . . . . . . . . . . . . . . . . . . . . *Libellula quadrimaculata*
[length of exuviae 21–27mm] **H**

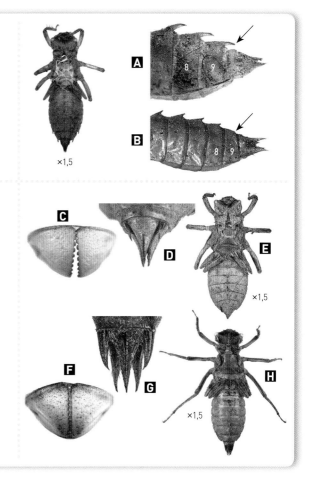

×1,5

# Genus *Orthetrum*

Lateral spines on segment 9 well developed. Length
of exuviae 21–29mm **A**.
. . . . . . . . . . . . . . . . . . . . . . . . . . . . . . . . . . .2

Lateral spines on segment 9 very small or absent **B**.
Length of exuviae <21mm.
. . . . . . . . . . . . . . . . . *Orthetrum coerulescens* **C** or
. . . . . . . . . . . . . . . . . . . . . *Orthetrum brunneum* **D**
[length of exuviae 16–21mm]

⚠ Note: For specific identification, use a binocular
microscope and to consult the works cited in the
bibliography on p. 147.

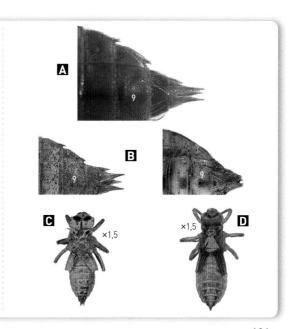

×1,5

**2.** Dorsal spines clearly visible on abdomen 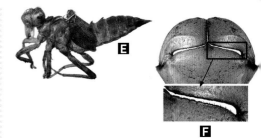.
Setae on mask between mentum and labial palps
short or absent **F**.
. . . . . . . . . . . . . . . . . . . . . . . . . . . . . . . . . . . . . .**3**

**2'.** No dorsal spines on abdomen **G**.
Setae between mentum and labial palps long **H**.
. . . . . . . . . . . . . . . . . . . . . . . . . . *Orthetrum albistylum*
[length of exuviae 21–25mm] **I**

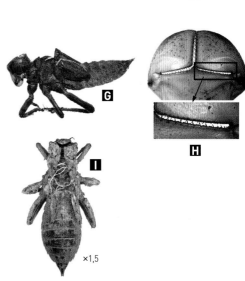

×1,5

**3.** Length of paraproct (L. para)/length of segment 9 (S9)
<1.5* **J** . . . . . . . . . . . . . . . . . *Orthetrum cancellatum*
[length of exuviae 21–29mm]

* To measure this character, it is important to exclude
the bundle of bristles at the end of the paraprocts.

**3'.** Length of paraproct (L. para)/length of segment 9
(S9) >1.5* **K** . . . . . . . . . . . . . . . .*Orthetrum trinacria*
[length of exuviae 24–28mm]

* To measure this character, it is important to exclude
the bundle of bristles present at the end of the
paraprocts.

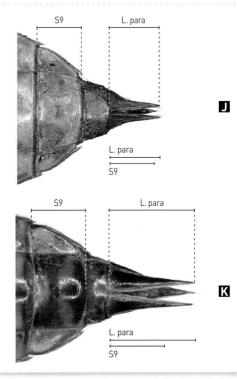

In dorsal view, lateral spines on segment 9 (S9) reach or exceed half the length of cerci 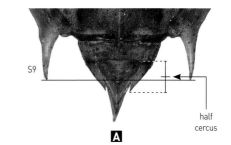.
. . . . . . . . . . . . . . . . . . . . . . . . *Brachythemis impartita*
[length of exuviae 14–18mm]

S9 / half cercus

**A**

. In dorsal view, lateral spines on segment 9 less than half the length of cerci **B**.
. . . . . . . . . . . . . . . . . . . . . . . . . . . . . . . . . . . . . . . . . . . .**2**

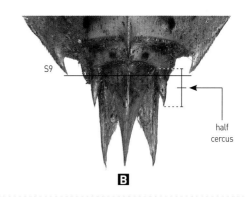

S9 / half cercus

**B**

Base of dorsal spine on segment 6 (B)/height of segment 6 (H) <1.5 **C**.
. . . . . . . . . . . . . . . . . . . . . . . . . . . . . .*Trithemis annulata*
[length of exuviae 16–20mm]

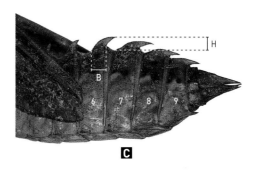

H / B / 6 7 8 9

**C**

. Base of dorsal spine on segment 6 (B)/height of segment 6 (H) >1.9 **D**.
. . . . . . . . . . . . . . . . . . . . . . . . . . . . . .*Trithemis kirbyi*
[length of exuviae 17–22mm]

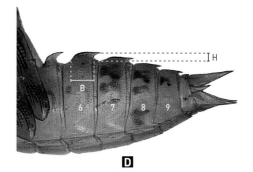

H / B / 6 7 8 9

**D**

Table 0

# Genus *Leucorrhinia*

**1.** Lateral spines present on segment 7 **A**.
In most cases, dorsal spine present on segment 9.
. . . . . . . . . . . . . . . . . . . . . . . . . . . *Leucorrhinia caudalis*
[length of exuviae 17–21mm]

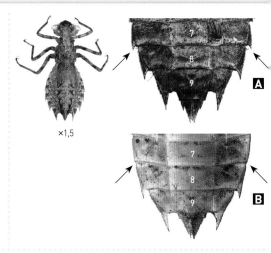

×1,5

**1'.** No lateral spines on segment 7 **B**.
No dorsal spine on segment 9.
. . . . . . . . . . . . . . . . . . . . . . . . . . . . . . . . . . . . .**2**

**2.** Lateral spines on segment 9 (LS9) more than the
length of segment 9 (S9) **C**.
. . . . . . . . . . . . . . . . . . . . . . . . . .*Leucorrhinia albifrons*
[length of exuviae 15–22mm]

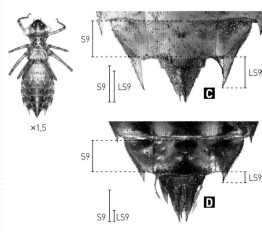

×1,5

**2'.** Lateral spines of segment 9 (LS9) less than half
the length of segment 9 **D**.
. . . . . . . . . . . . . . . . . . . . . . . . . . . . . . . . . . . .**3**

**3.** Dorsal spines present up to segment 8 **E**.
. . . . . . . . . . . . . . . . . . . . . . . . . .*Leucorrhinia pectoralis*
[length of exuviae 19–23mm]

**3'.** No dorsal spines on segment 8 **F**.
. . . . . . . . . . . . . . . . . . . . . . . . . .*Leucorrhinia dubia* **G**
or *Leucorrhinia rubicunda* **H**
[length of exuviae 16–22mm]

⚠ Note: For specific identification, use a binocular
microscope and consult the works cited in the
bibliography on p. 147.

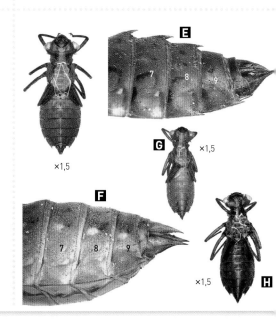

×1,5

No dorsal spines on abdomen **A**.
. . . . . . . . . . . . . . . . . . . . . . . . . . . . . . . . . . . . . . . . . . . . . . . . . . . **2**

**A**

'. Dorsal spines present on abdomen **B**.
. . . . . . . . . . . . . Genera *Selysiothemis* and *Sympetrum*
excluding *Sympetrum fonscolombii*

⚠ Note: *Sympetrum* (except *Sympetrum fonscolombii*) and *Selysiothemis nigra* cannot be identified with a magnifying glass in the field. For specific identification, use a binocular microscope and consult the works cited in the bibliography on p. 147.

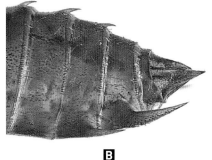

**B**

Cerci more than half the length of paraprocts **C**.
. . . . . . . . . . . . . . . . . . . . . . . . . . . *Crocothemis erythraea*
[length of exuviae 17–19mm]

×1,5

paraprocts · cerci

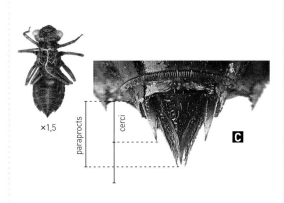

**C**

'. Cerci less than half the length of paraprocts **D**.
. . . . . . . . . . . . . . . . . . . . . . . *Sympetrum fonscolombii*
[length of exuviae 15–20mm]

⚠ Note: If in doubt, use a binocular microscope and consult the references cited in the bibliography on p. 147.

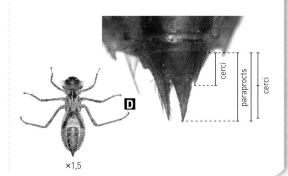

**D**

cerci · paraprocts · cerci

×1,5

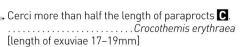

Banded Demoiselle, *Calopteryx splendens*

Emerald Damselfly, *Lestes sponsa*

# Pterography (Wing photography)

Included here are black-and-white or greyscale images of the male wing, and of the female wing when this differs from the male, for the various families and species covered in this book. The images have been produced by scanning wings at a resolution of 2,400ppi (pixels per inch). They give an accurate representation of the wing venation of adult dragonflies, an important character in the identification of families, genera and sometimes even species. Unless otherwise specified (for small species), wings are shown at actual size.

## Zygoptera

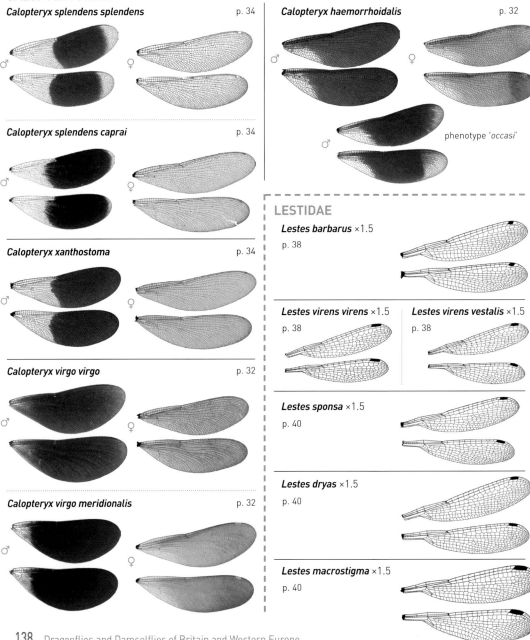

**CALOPTERYGIDAE**

*Calopteryx splendens splendens* — p. 34

*Calopteryx splendens caprai* — p. 34

*Calopteryx xanthostoma* — p. 34

*Calopteryx virgo virgo* — p. 32

*Calopteryx virgo meridionalis* — p. 32

*Calopteryx haemorrhoidalis* — p. 32

phenotype *'occasi'*

**LESTIDAE**

*Lestes barbarus* ×1.5
p. 38

*Lestes virens virens* ×1.5
p. 38

*Lestes virens vestalis* ×1.5
p. 38

*Lestes sponsa* ×1.5
p. 40

*Lestes dryas* ×1.5
p. 40

*Lestes macrostigma* ×1.5
p. 40

**LESTIDAE** continued

*Chalcolestes viridis* ×1.5 p. 38

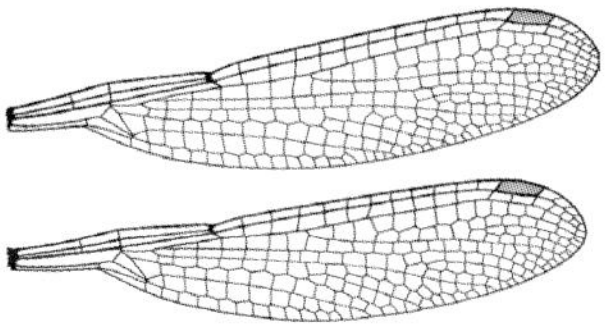

*Chalcolestes parvidens* ×1.5 p. 38

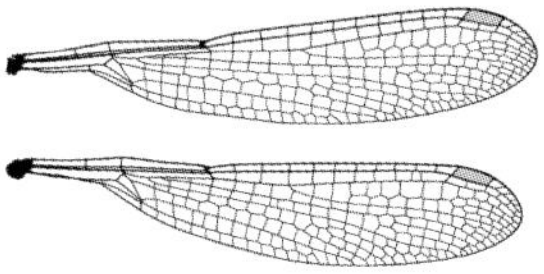

*Sympecma fusca* ×1.5 p. 36

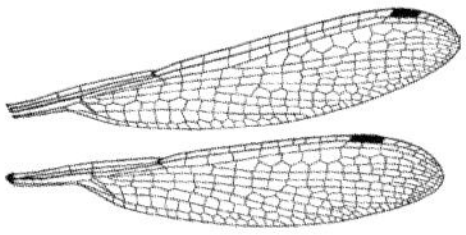

*Sympecma paedisca* ×1.5 p. 36

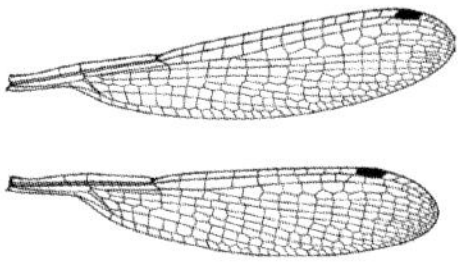

**PLATYCNEMIDIDAE**

*Platycnemis pennipes* ×1.5 p. 42

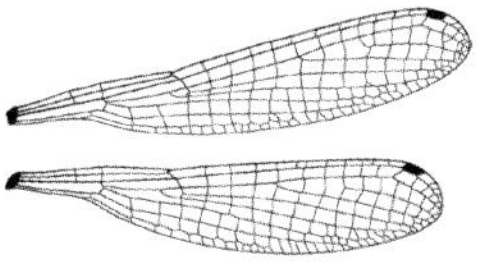

*Platycnemis latipes* ×1.5 p. 42

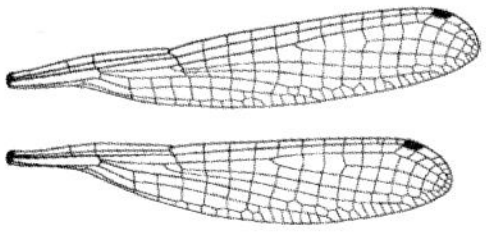

*Platycnemis acutipennis* ×1.5 p. 42

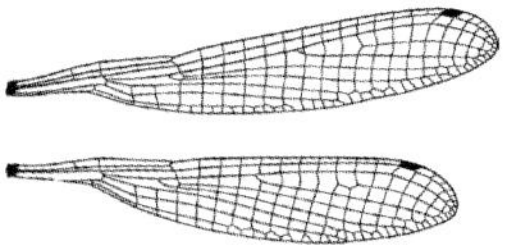

**COENAGRIONIDAE**

*Coenagrion puella* ×1.5 p. 52

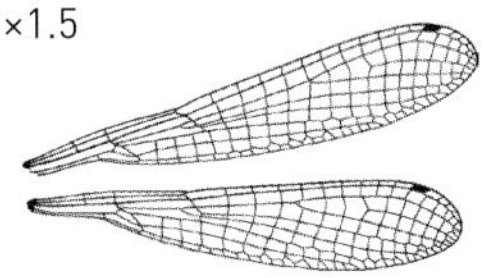

*Coenagrion pulchellum* ×1.5 p. 52

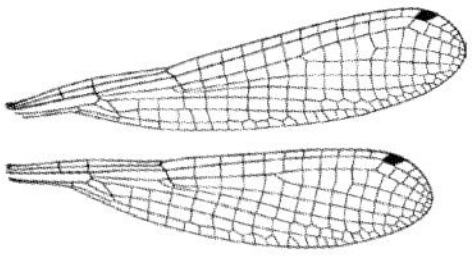

*Coenagrion hastulatum* ×1.5 p. 50

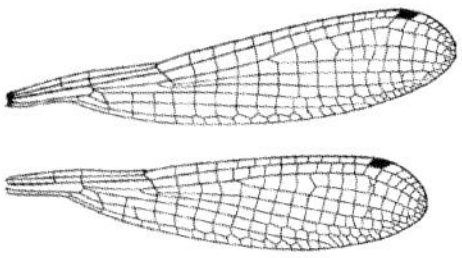

*Coenagrion lunulatum* ×1.5 p. 48

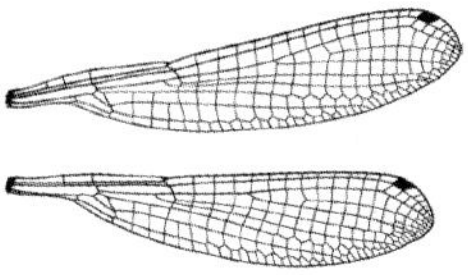

*Coenagrion ornatum* ×1.5 p. 52

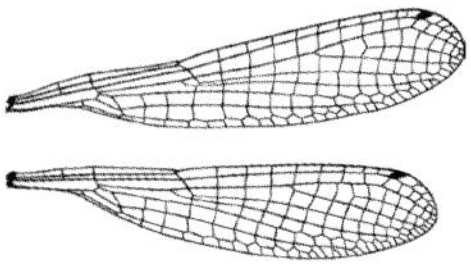

*Coenagrion mercuriale* ×1.5 p. 48

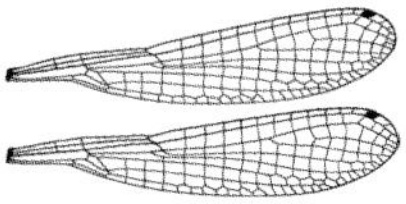

*Coenagrion scitulum* ×1.5 p. 50

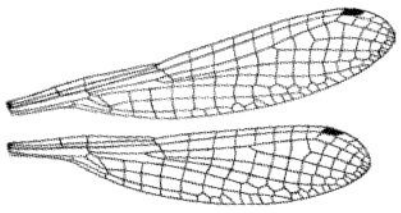

*Coenagrion caerulescens* ×1.5 p. 50

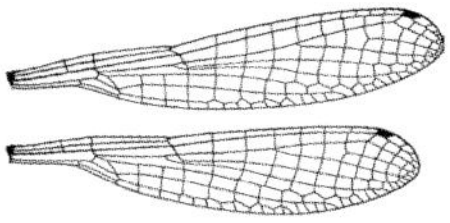

*Enallagma cyathigerum* ×1.5        p. 44

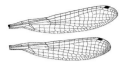

*Erythromma lindenii* ×1.5        p. 56

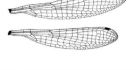

*Ischnura elegans* ×1.5        p. 54

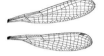

*Erythromma najas* ×1.5        p. 56

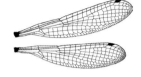

*Ischnura genei* ×1.5        p. 54

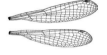

*Erythromma viridulum* ×1.5        p. 56

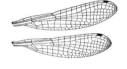

*Ischnura graellsii* ×1.5        p. 54

*Pyrrhosoma nymphula* ×1.5        p. 46

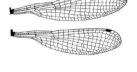

*Ischnura pumilio* ×1.5        p. 54

*Ceriagrion tenellum* ×1.5        p. 46

*Nehalennia speciosa* ×1.5        p. 44

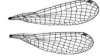

# Anisoptera
## AESHNIDAE

**Aeshna affinis**    p. 64

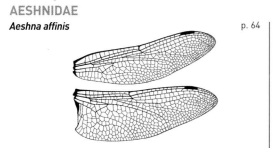

**Aeshna caerulea**    p. 60

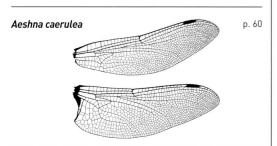

**Aeshna mixta**    p. 64

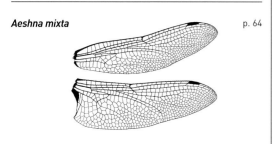

**Aeshna juncea**    p. 62

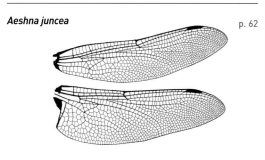

**Aeshna subarctica**    p. 62

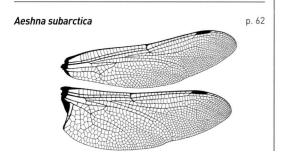

**Aeshna cyanea**    p. 62

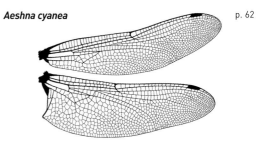

**Aeshna grandis**    p. 60

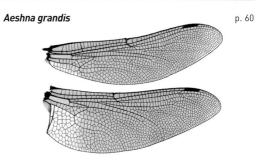

**Aeshna isoceles**    p. 60

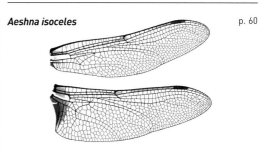

**Brachytron pratense**    p. 64

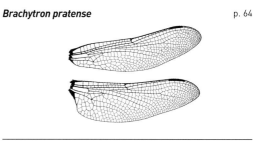

**Boyeria irene**    p. 58

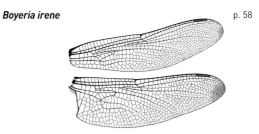

**AESHNIDAE** continued

*Hemianax ephippiger*  p. 58

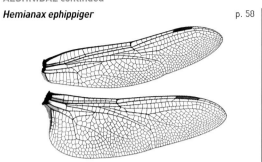

*Anax junius*  p. 66

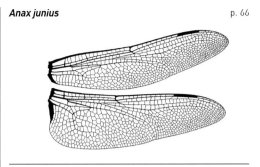

*Anax imperator*  p. 66

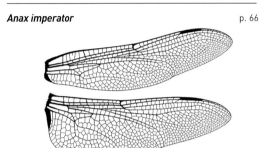

*Anax parthenope*  p. 66

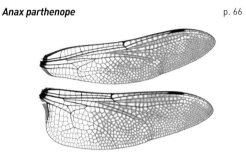

## GOMPHIDAE

*Gomphus vulgatissimus*  p. 70

*Gomphus graslinii*  p. 70

*Onychogomphus forcipatus forcipatus*

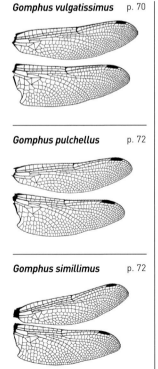

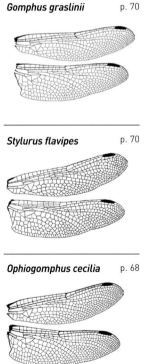

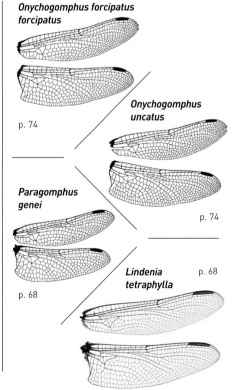

*Gomphus pulchellus*  p. 72

*Stylurus flavipes*  p. 70

p. 74

*Onychogomphus uncatus*

*Gomphus simillimus*  p. 72

*Ophiogomphus cecilia*  p. 68

*Paragomphus genei*

p. 74

p. 68

*Lindenia tetraphylla*  p. 68

## CORDULEGASTRIDAE

**Cordulegaster boltonii**  p. 76

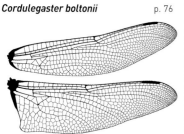

**Cordulegaster bidentata**  p. 76

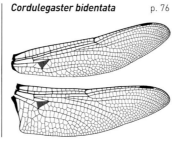

## MACROMIIDAE

**Macromia splendens**  p. 76

## CORDULIIDAE and *OXYGASTRA CURTISII*

**Cordulia aenea**  p. 78

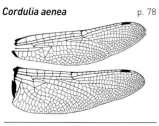

**Somatochlora flavomaculata**  p. 80

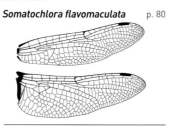

**Oxygastra curtisii**  p. 78

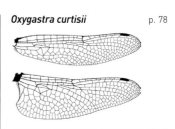

**Somatochlora metallica**  p. 80

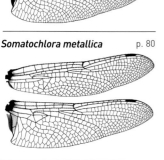

**Somatochlora alpestris**  p. 82

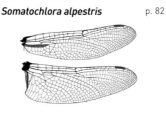

**Epitheca bimaculata**  p. 78

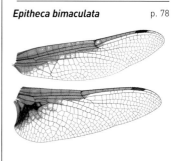

**Somatochlora meridionalis**  p. 80

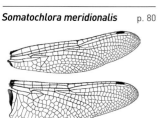

**Somatochlora arctica**  p. 82

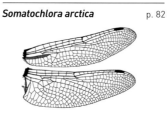

## LIBELLULIDAE

**Libellula quadrimaculata**  p. 88

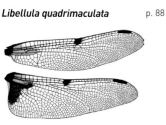

**Libellula depressa**  p. 88

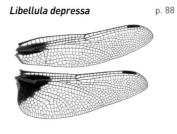

**Libellula fulva**  p. 88

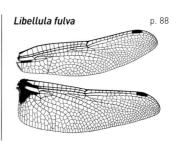

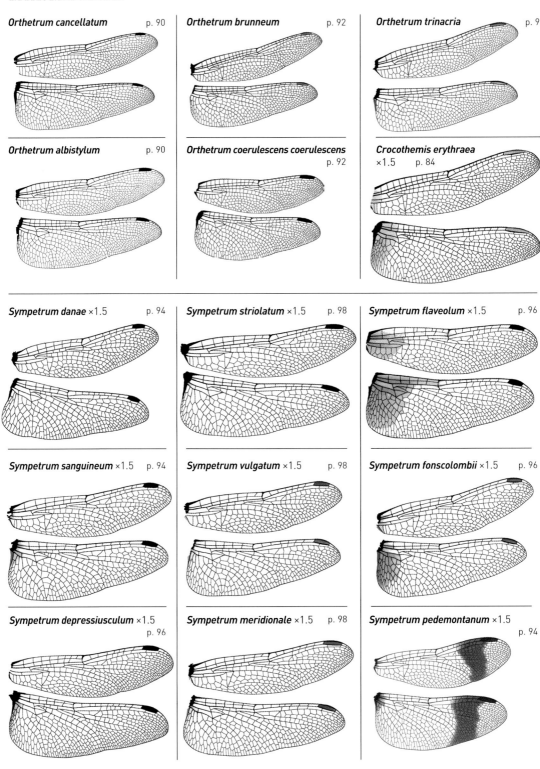

**Orthetrum cancellatum** p. 90

**Orthetrum brunneum** p. 92

**Orthetrum trinacria** p. 9[

**Orthetrum albistylum** p. 90

**Orthetrum coerulescens coerulescens** p. 92

**Crocothemis erythraea** ×1.5 p. 84

**Sympetrum danae** ×1.5 p. 94

**Sympetrum striolatum** ×1.5 p. 98

**Sympetrum flaveolum** ×1.5 p. 96

**Sympetrum sanguineum** ×1.5 p. 94

**Sympetrum vulgatum** ×1.5 p. 98

**Sympetrum fonscolombii** ×1.5 p. 96

**Sympetrum depressiusculum** ×1.5 p. 96

**Sympetrum meridionale** ×1.5 p. 98

**Sympetrum pedemontanum** ×1.5 p. 94

*Leucorrhinia dubia* ×1.5     p. 102

*Leucorrhinia rubicunda* ×1.5     p. 102

*Leucorrhinia pectoralis* ×1.5     p. 100

*Leucorrhinia caudalis* ×1.5     p. 100

*Leucorrhinia albifrons* ×1.5     p. 100

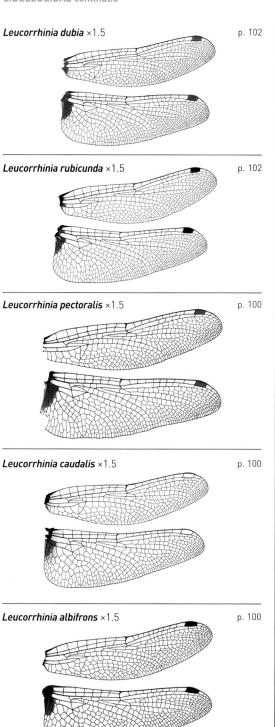

*Trithemis annulata* ×1.5     p. 86

*Brachythemis impartita* ×1.5     p. 84

*Selysiothemis nigra* ×1.5     p. 86

*Pantala flavescens* ×1.5     p. 84

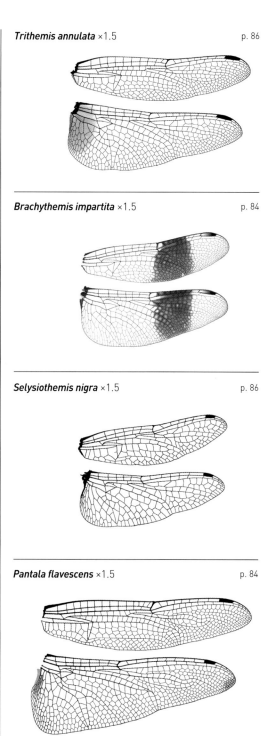

# Comparative Plate of Dragonfly Morphology

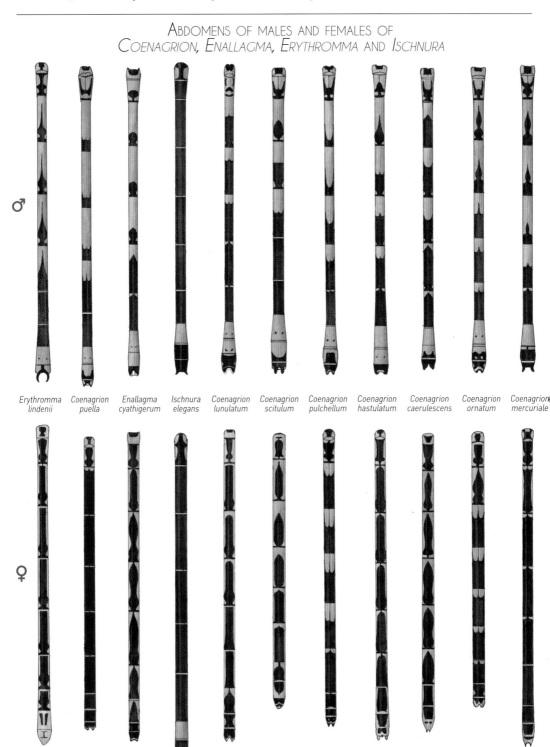

♂

Erythromma
lindenii

Coenagrion
puella

Enallagma
cyathigerum

Ischnura
elegans

Coenagrion
lunulatum

Coenagrion
scitulum

Coenagrion
pulchellum

Coenagrion
hastulatum

Coenagrion
caerulescens

Coenagrion
ornatum

Coenagrion
mercuriale

♀

Broad-bodied Chaser, *Libellula depressa*

# Checklist of the Dragonflies and Damselflies of Britain and Western Europe

## HOW TO READ THIS LIST

In the list below, the countries treated in this guide are indicated by the following letters: F = continental France; Co = Corsica; Be = Belgium; N = Netherlands; L = Luxembourg; S = Switzerland; GB = Great Britain; I = Ireland. A hash (#) indicates that the species is no longer considered to be breeding in the country, although isolated erratic individuals are still observed from time to time. A question mark (?) indicates that records for the species for the country are questionable. Square brackets indicate that observations relate only to individuals deemed to be erratic or migratory. When no subspecies is mentioned, the information provided always refers to the nominate form. The vernacular names used in this list follow Dijkstra & Lewington (2020).

Ec: species of flowing waters

Es: species of stagnant waters

M: migratory species

IUCN (International Union for Conservation of Nature): endangered species that are included in Red Lists of threatened species, regularly updated by panels of experts. These lists are particularly useful because they are based on objective criteria. They can help alert the international community but do not impose any constraints. They are fully independent of lists that can be found in European and national legislation.

BC: species in the Bern Convention, ratified on 19 September 1979 by 19 member states of the Council of Europe.

AnII: species in Annex II to the Directive 92/43/EEC, also known as the Habitats Directive, amended by Directive 97/62/EC of 27 October 1997, by the Treaty of Accession of the 10 new member states of the European Union on 23 September 2003, and by Council Directive 2013/17/EU of 13 May 2013 adapting certain directives in the field of environment, by reason of the accession of the Republic of Croatia. It aims to protect certain natural habitats, and the most threatened species at European level.

AnIV: species in Annex IV to the Directive 92/43/EEC, also known as the Habitats Directive; species of community interest in need of strict protection. They need to be protected regardless of their habitat, and unauthorised capture is prohibited.

# Odonata Fabricius, 1793

## Zygoptera Selys, 1854

### Calopterygidae Selys, 1850
**Calopteryx** Leach, 1815
1. *Calopteryx splendens splendens* (Harris, 1780) (Banded Demoiselle) . . . . . . . . **Ec**, F Be N L S GB I
2. *Calopteryx splendens caprai* (Conci, 1956) (Banded Demoiselle). . . . . . . . . . . . **Ec**, Co N S GB I
3. *Calopteryx xanthostoma* (Charpentier, 1825) (Western Demoiselle) . . . . . . . . . . . . . **Ec**, F S?
4. *Calopteryx virgo virgo* (Linnaeus, 1758) (Beautiful Demoiselle) . . . . . . . . . **Ec**, F Be N L S GB I
5. *Calopteryx virgo meridionalis* (Selys, 1873) (Beautiful Demoiselle) . . . . . . . **Ec**, F Co N S GB I
6. *Calopteryx haemorrhoidalis* (Vander Linden, 1825) (Copper Demoiselle) . . . . . . . . . . . **Ec**, F Co

### Lestidae Calvert, 1901
**Lestes** Leach, 1815
7. *Lestes barbarus* (Fabricius, 1798) (Southern Emerald Damselfly; Migrant Spreadwing) . . . **Es**, F Co Be N L S GB
8. *Lestes virens virens* (Charpentier, 1825) (Small Spreadwing). . . . . . . . . . . . . . **Es**, F Co N
9. *Lestes virens vestalis* (Rambur, 1842) (Small Spreadwing) . . . . . . . . . . . . . **Es**, F Be N L S
10. *Lestes sponsa* (Hansemann, 1823) (Emerald Damselfly; Common Spreadwing) . . . **Es**, F Be N L S GB I
11. *Lestes dryas* (Kirby, 1890) (Scarce Emerald Damselfly; Robust Spreadwing) . . . **Es**, F Be N L S GB I
12. *Lestes macrostigma* (Eversmann, 1836) (Dark Spreadwing) . . . . . . . . . **Es**, F Co [S]?
**Chalcolestes** Kennedy, 1920
13. *Chalcolestes viridis* (Vander Linden, 1825) (Willow Emerald Damselfly) . . . . . **Ec**, **Es**, F Co Be L S GB
14. *Chalcolestes parvidens* (Artobolevskij, 1929) (Eastern Willow Spreadwing) . . . . . . . **Ec**, **Es**, Co
**Sympecma** Burmeister, 1839
15. *Sympecma fusca* (Vander Linden, 1820) (Common Winter Damselfly) . . . . **Es**, F Co Be N L S GB
16. *Sympecma paedisca* (Brauer, 1877) (Siberian Winter Damselfly) . . . . . . . . . . . **Es**, **BC**, **AnIV**, F# N S

### Platycnemididae Jacobson and Bianchi, 1905
**Platycnemis** Burmeister, 1839
17. *Platycnemis pennipes* (Pallas, 1771) (White-legged Damselfly; Blue Featherleg) . . **Ec**, **Es**, F Co? Be N L S GB
18. *Platycnemis latipes* (Rambur, 1842) (White Featherleg). . . . . . . . . . . . . . . . . . **Ec**, F
19. *Platycnemis acutipennis* (Selys, 1841) (Orange White-legged Damselfly; Orange Featherleg) . **Ec**, **Es**, F

### Coenagrionidae Kirby, 1890
**Coenagrion** Kirby, 1890
20. *Coenagrion puella* (Linnaeus, 1758) (Azure Bluet; Azure Damselfly) . . . . **Es**, F Co Be N L S GB I
21. *Coenagrion pulchellum* (Vander Linden, 1825) (Variable Bluet; Variable Damselfly) . . . **Es**, F Co Be N L# S GB I

22. *Coenagrion hastulatum* (Charpentier, 1825) (Spearhead Bluet; Northern Damselfly). . . **Es**, F Be N S GB

23. *Coenagrion lunulatum* (Charpentier, 1840) (Crescent Bluet; Irish Damselfly) . . . . . . **Es**, F Be N S# GB I

24. *Coenagrion ornatum* (Selys, 1850) (Ornate Bluet) . . . . . . . . . . . . . . **Ec, AnII**, F S#

25. *Coenagrion mercuriale* (Charpentier, 1840) (Mercury Bluet; Southern Damselfly) . **Ec, IUCN, BC, AnII**, F Be N L S GB

26. *Coenagrion scitulum* (Rambur, 1842) (Dainty Bluet; Dainty Damselfly). . . . **Es**, F Co Be N L S GB

27. *Coenagrion caerulescens* (Fonscolombe, 1838) (Mediterranean Bluet) . . . . . . . . . **Ec**, F Co

**Enallagma** Charpentier, 1840

28. *Enallagma cyathigerum* (Charpentier, 1840) (Common Bluet; Common Blue Damselfly). . . **Es**, F Co Be N L S GB I

**Ischnura** Charpentier, 1840

29. *Ischnura elegans* (Vander Linden, 1820) (Common Bluetail; Blue-tailed Damselfly) . . . **Es**, F Be N L S GB I

30. *Ischnura genei* (Rambur, 1842) (Island Bluetail). . . . . . . . . . . . . . . **Es**, Co

31. *Ischnura graellsii* (Rambur, 1842) (Iberian Bluetail). . . . . . . . . . . . . . . . **Es**, F

32. *Ischnura pumilio* (Charpentier, 1825) (Scarce Blue-tailed Damselfly) . . . . . **Es**, F Be N L S GB I

**Erythromma** Charpentier, 1840

33. *Erythromma lindenii* (Selys, 1840) (Blue-eye; Goblet-marked Damselfly) . . . . . .**Ec**, **Es**, F Co Be N L S

34. *Erythromma najas* (Hansemann, 1823) (Large Redeye; Red-eyed Damselfly) . . . . . **Es**, F Be N L S GB

35. *Erythromma viridulum* (Charpentier, 1840) (Small Redeye; Small Red-eyed Damselfly). . . . **Es**, F Co Be N L S GB

**Pyrrhosoma** Charpentier, 1840

36. *Pyrrhosoma nymphula* (Sulzer, 1776) (Large Red Damselfly) . . . . . . . **Es**, F Co? Be N L S GB I

**Ceriagrion** Selys, 1876

37. *Ceriagrion tenellum* (Villers, 1789) (Small Red Damselfly) . . . . . . . . . **Ec**, **Es**, F Co Be N S GB

**Nehalennia** Selys, 1850

38. *Nehalennia speciosa* (Charpentier, 1840) (Sedgling; Pygmy Damselfly) . . . . . . . **Es**, **IUCN**, F Be# N L# S


**Anisoptera** Selys, 1854

**Aeshnidae** Leach, 1815

**Aeshna** Fabricius, 1775

39. *Aeshna affinis* (Vander Linden, 1820) (Blue-eyed Hawker; Southern Migrant Hawker) . . **Es**, F Co Be N L S GB

40. *Aeshna caerulea* (Ström, 1783) (Azure Hawker) . . . . . . . . . . . . . . **Es**, F S GB

41. *Aeshna mixta* (Latreille, 1805) (Migrant Hawker) . . . . . . . . . . . . **Es**, F Co Be N L S GB I

42. *Aeshna juncea* (Linnaeus, 1758) (Moorland Hawker; Common Hawker) . . . **Es**, F Be N L# S GB I

43. *Aeshna subarctica* (Djakonov, 1922) (Bog Hawker; Subarctic Hawker) . . . . . . . . . . . . **Es**, F Be N S

44. *Aeshna cyanea* (Müller, 1764) (Blue Hawker; Southern Hawker) . . . . . . **Es**, F Co Be N L S GB I

45. *Aeshna grandis* (Linnaeus, 1758) (Brown Hawker) . . . . . . . . . . . . . . **Es**, F Be N L S GB I

46. *Aeshna isoceles* (Müller, 1767) (Green-eyed Hawker; Norfolk Hawker) . . . **Es**, F Co Be N L S GB

**Brachytron** Evans, 1845
47. *Brachytron pratense* (Müller, 1764) (Hairy Hawker; Hairy Dragonfly). . . . **Es**, F Co Be N L S GB I

**Boyeria** McLachlan, 1896
48. *Boyeria irene* (Fonscolombe, 1838) (Western Spectre; Dusk Hawker) . . . . . . . **Ec**, **[Es]**, F Co S

**Hemianax** Selys, 1883
49. *Hemianax ephippiger* (Burmeister, 1839) (Vagrant Emperor) . . . . . . . **Es**, **M**, F Co [Be] N S GB I

**Anax** Leach, 1815
50. *Anax imperator* (Leach, 1815) (Blue Emperor; Emperor Dragonfly) . . . . **Es**, F Co Be N L S GB I
51. *Anax junius* (Drury, 1773) (Green Darner) . . . . . . . . . . . . . . . . **M**, [F] GB
52. *Anax parthenope* (Selys, 1839) (Lesser Emperor) . . . . . . . . . . . . **Es**, F Co Be N L S GB I


## **Gomphidae** Rambur, 1842
**Gomphus** Leach, 1815
53. *Gomphus vulgatissimus* (Linnaeus, 1758) (Common Clubtail; Club-tailed Dragonfly). . . **Ec**, **[Es]**,
    F Be N L S GB I
54. *Gomphus pulchellus* (Selys, 1840) (Western Clubtail) . . . . . . . . . . . . . **Ec**, **Es**, F Be N L S
55. *Gomphus simillimus* (Selys, 1840) (Yellow Clubtail) . . . . . . . . . . . . . . . . **Ec**, F Be S
56. *Gomphus graslinii* (Rambur, 1842) (Pronged Clubtail). . . . . . . . . **Ec**, **IUCN**, **BC**, **AnII**, **AnIV**, F

**Stylurus** Needham, 1897
57. *Stylurus flavipes* (Charpentier, 1825) (River Clubtail) . . . . . . . . .**Ec**, **BC**, **AnIV**, F Be N L# S

**Ophiogomphus** Selys, 1854
58. *Ophiogomphus cecilia* (Geoffroy in Fourcroy, 1785) (Green Snaketail; Green Clubtail) . . . **Ec**, **BC**,
    **AnII**, **AnIV**, F N L S

**Onychogomphus** Selys, 1854
59. *Onychogomphus forcipatus forcipatus* (Linnaeus, 1758) (Small Pincertail; Green-eyed
    Hooktail) . . . . . . . . . . . . . . . . . . . . . . . . . . . . . . . . **Ec**, **[Es]**, F Be N L S
60. *Onychogomphus forcipatus unguiculatus* (Vander Linden, 1820) (Small Pincertail; Green-eyed
    Hooktail) . . . . . . . . . . . . . . . . . . . . . . . . . . . . . . . . . . . . . **Ec**, F S
61. *Onychogomphus uncatus* (Charpentier, 1840) (Large Pincertail; Blue-eyed
    Hooktail) . . . . . . . . . . . . . . . . . . . . . . . . . . . . . . . . . . . .**Ec**, F [Be] S

**Paragomphus** Cowley, 1934
62. *Paragomphus genei* (Selys, 1841) (Green Hooktail) . . . . . . . . . . . . . . . . **Ec**, **Es**, Co

**Lindenia** de Haan, 1826
63. *Lindenia tetraphylla* (Vander Linden, 1825) (Bladetail) . . . . . . . . . . . **Es**, **BC**, **AnII**, **AnIV**, Co


## **Cordulegastridae** Hagen, 1875
**Cordulegaster** Leach, 1815
64. *Cordulegaster boltonii boltonii* (Donovan, 1807) (Common Goldenring; Golden-ringed
    Dragonfly) . . . . . . . . . . . . . . . . . . . . . . . . . . . . . .**Ec**, F [Co] Be N L S GB I
65. *Cordulegaster boltonii immaculifrons* (Selys, 1850) (Common Goldenring; Golden-ringed
    Dragonfly) . . . . . . . . . . . . . . . . . . . . . . . . . . . . . . . . . . **Ec**, F N GB I
66. *Cordulegaster bidentata* (Selys, 1843) (Sombre Goldenring; Two-toothed Goldenring) **Ec**, F Be L S


## **Macromiidae** Needham, 1903
**Macromia** Rambur, 1842
67. *Macromia splendens* (Pictet, 1843) (Splendid Cruiser) . . . . . . . . . **Ec**, **IUCN**, **BC**, **AnII**, **AnIV**, F

**Corduliidae** Selys, 1850

*Cordulia* Leach, 1815

68. *Cordulia aenea* (Linnaeus, 1758) (Downy Emerald) . . . . . . . . . . . . **Es**, F Be N L S GB I

*Somatochlora* Selys, 1871

69. *Somatochlora metallica* (Vander Linden, 1825) (Brilliant Emerald) . . . . . . . . **Es**, F Be N L S GB

70. *Somatochlora meridionalis* (Nielsen, 1935) (Balkan Emerald) . . . . . . . . . . . . . **Ec**, F Co

71. *Somatochlora flavomaculata* (Vander Linden, 1825) (Yellow-spotted Emerald) . . . . . **Es**, F Co Be N L# S

72. *Somatochlora alpestris* (Selys, 1840) (Alpine Emerald) . . . . . . . . . . . . . . . . . **Es**, F S

73. *Somatochlora arctica* (Zetterstedt, 1840) (Northern Emerald) . . . . . . . . . **Es**, F Be N S GB I

*Epitheca* Burmeister, 1839

74. *Epitheca bimaculata* (Charpentier, 1825) (Eurasian Baskettail; Two-spotted Dragonfly) . . . **Es**, F Be L S

**Incertae sedis**

*Oxygastra* Selys, 1870

75. *Oxygastra curtisii* (Dale, 1834) (Orange-spotted Emerald) . . . . . **Ec**, [**Es**], **IUCN**, **BC**, **AnII**, **AnIV**, F Be N L S GB

**Libellulidae** Leach, 1815

*Libellula* Linnaeus, 1758

76. *Libellula quadrimaculata* (Linnaeus, 1758) (Four-spotted Chaser) . . . . . **Es**, F Co Be N L S GB I

77. *Libellula depressa* (Linnaeus, 1758) (Broad-bodied Chaser) . . . . . . . **Es**, F Co Be N L S GB I

78. *Libellula fulva* (Müller, 1764) (Scarce Chaser) . . . . . . . . . . . . . . . . **Es**, F Co Be N L S GB I

*Orthetrum* Newman, 1833

79. *Orthetrum cancellatum* (Linnaeus, 1758) (Black-tailed Skimmer) . . . . . **Es**, F Co Be N L S GB I

80. *Orthetrum albistylum* (Selys, 1848) (White-tailed Skimmer) . . . . . . . . . . . **Es**, F Be S

81. *Orthetrum brunneum* (Fonscolombe, 1837) (Southern Skimmer) . . . . . . **Ec**, **Es**, F Co Be N L S

82. *Orthetrum coerulescens coerulescens* (Fabricius, 1798) (Keeled Skimmer) . . . . .**Ec**, **Es**, F Co Be N L S GB I

83. *Orthetrum coerulescens anceps* (Schneider, 1845) (Keeled Skimmer) . . . . . . . **Ec**, **Es**, Co N GB I

84. *Orthetrum trinacria* (Selys, 1841) (Long Skimmer) . . . . . . . . . . . . . . . . . . **Es**, Co

*Crocothemis* Brauer, 1868

85. *Crocothemis erythraea* (Brullé, 1832) (Broad Scarlet; Scarlet Darter) . . . . **Es**, F Co Be N L S GB

*Sympetrum* Newman, 1833

86. *Sympetrum danae* (Sulzer, 1776) (Black Darter) . . . . . . . . . . . . . . **Es**, F Be N L S GB I

87. *Sympetrum sanguineum* (Müller, 1764) (Ruddy Darter) . . . . . . . . . . **Es**, F Co Be N L S GB I

88. *Sympetrum depressiusculum* (Selys, 1841) (Spotted Darter; Marshland Darter) . . . . . . . . **Es**, F Co? Be N L# S

89. *Sympetrum vulgatum vulgatum* (Linnaeus, 1758) (Vagrant Darter; Moustached Darter) . . . **Es**, F Be N L S GB

90. *Sympetrum vulgatum ibericum* (Ocharan, 1985) (Vagrant Darter; Moustached Darter) . **Es**, F N GB

91. *Sympetrum striolatum* (Charpentier, 1840) (Common Darter) . . . . . **Ec**, **Es**, F Co Be N L S GB I

92. *Sympetrum meridionale* (Selys, 1841) (Southern Darter) . . . . . . . . . **Es**, F Co Be N L S GB

93. *Sympetrum flaveolum* (Linnaeus, 1758) (Yellow-winged Darter) . . . . . . . . **Es**, F Be N L S GB I

94. Sympetrum fonscolombii (Selys, 1840) (Red-veined Darter) . . . . . . . . **Es**, F Co Be N L S GB I

95. Sympetrum pedemontanum (Müller in Allioni, 1766) (Banded Darter) . [Ec], Es, F Be N L# S GB

**Leucorrhinia** Brittinger, 1850

96. *Leucorrhinia dubia* (Vander Linden, 1825) (Small Whiteface; White-faced Darter) . . . Es, F Be N L S GB

97. *Leucorrhinia rubicunda* (Linnaeus, 1758) (Ruby Whiteface) . . . . . . . . . . . Es, F# Be N L [S]

98. *Leucorrhinia pectoralis* (Charpentier, 1825) (Yellow-spotted Whiteface; Large White-faced Darter) . . . . . . . . . . . . . . . . . . . . . . . . . . . Es, BC, AnII, AnIV, F Be N L S GB

99. *Leucorrhinia caudalis* (Charpentier, 1840) (Lilypad Whiteface; Dainty White-faced Darter) . . . Es, BC, AnIV, F Be N L S

100. *Leucorrhinia albifrons* (Burmeister, 1839) (Dark Whiteface; Eastern White-faced Darter) . . . Es, BC, AnIV, F N S

**Trithemis** Brauer, 1868

101. *Trithemis annulata* (Palisot de Beauvois, 1807) (Violet Dropwing; Violet-marked Darter) . . . Ec, Es, F Co

102. *Trithemis kirbyi* (Selys, 1891) (Orange-winged Dropwing) . . . . . . . . . . . . . . [F]

**Brachythemis** Brauer, 1868

103. *Brachythemis impartita* (Karsch, 1890) (Northern Banded Groundling) . . . . . . . . [Ec], Es, Co

**Pantala** Hagen, 1861

104. *Pantala flavescens* (Fabricius, 1798) (Globe Skimmer; Globe Wanderer; Wandering Glider) . Es, [F]? GB

**Selysiothemis** Ris, 1897

105. *Selysiothemis nigra* (Vander Linden, 1825) (Black Pennant) . . . . . . . . . . . . . . . . Co

# Authors

**Jean-Pierre Boudot**, PhD in Natural Sciences, is a researcher at the Centre national de la recherche scientifique (CNRS, French National Centre for Scientific Research), and has been studying dragonflies since the 1980s. He has authored and co-authored several publications on Odonata. Over the past decades, he has travelled extensively in Europe, North Africa and the Middle East to catalogue dragonfly species, and he has become one of the leading specialists on this group. A new dragonfly species discovered in 2014 in Morocco was named in his honour. A referee for the International Union for the Conservation of Nature (IUCN), he has been involved in the development of several Red Lists for Odonata. He is one of the major contributors to the *Atlas of the European Dragonflies and Damselflies*, published in English at the end of 2014.

**Guillaume Doucet** has been studying dragonflies for 15 years. His survey work in the Limousin region, Franche-Comté, south-west France, the Alps and Corsica has allowed him to gain in-depth knowledge of the species in this guide, particularly with regards to larval ecology. He subsequently created an identification key to the exuviae of metropolitan France, published in 2010 by the French Odonatological Society. In 2014, he contributed to the first edition of the *Field Guide to the Dragonflies of France, Belgium, Luxembourg and Switzerland*. In addition, he organises regular training sessions for naturalists who are keen to further their knowledge in the identification of larvae and exuviae.

**Daniel Grand** worked for the Greater Lyon Water Department and dedicated his life to dragonflies. He has authored and co-authored numerous publications and books, including the *Dragonflies of France, Belgium and Luxembourg*, written in collaboration with Jean-Pierre Boudot in 2006, and the *Field Guide to the Dragonflies of France, Belgium, Luxembourg and Switzerland*, written with Jean-Pierre Boudot and Guillaume Doucet in 2014.

# Acknowledgements

The authors would like to thank Stéphane Hette for his photographs, used with superb effect throughout this book.

Special thanks go to Madame Roberte Grand, who allowed the authors to pursue the work started by Daniel.

The authors would also like to thank Yves Doux, for his detailed illustrations.

Bloomsbury Wildlife would like to thank Marc Heath, Marianne Taylor and Oliver Wright for providing additional photos.

# Index of Common Names

# Index of Scientific Names

For the taxa listed in this index, **page numbers in bold** refer to adult individuals; other page numbers refer to larvae and exuviae. For habitat diagram, see pp20–21.